FAMILY

AGRICULTURE

D0813270

Tax inequities, federal subsidies of corporate farming, competition from foreign farms, the monopoly practices of grain companies, the fact that across America farming had become big business while on my father's farm it had not – all of this finally was forcing his hand ... No matter what, it was going to take an assumption of risk and some of Heinrich [Kohn's] original pioneer daring.

from *The Last Farmer: an American Memoir*,
by Howard Kohn, p 53

We use what we need from the land and try to save what we can and build up for the young ones, that's the right way, isn't it?
Marilena de Oliveira Kersting, a farmer and mother in
Southern Brazil

When I grow up? I don't know, but my family will tell me ...
Jesse N K Zinnah, a fifth grader in a rural school in Liberia

FAMILY AGRICULTURE

TRADITION AND
TRANSFORMATION

DAVID G FRANCIS

WITHDRAWN

EARTHSCAN
Earthscan Publications Ltd, London

Todd Wehr
Memorial Library

For the new generation of family agriculturalists.
You can do it.

First published in 1994 by
Earthscan Publications Limited
120 Pentonville Road, London N1 9JN

Copyright © David Francis, 1994

All rights reserved

A catalogue record for this book is available from the British Library

ISBN: 1 85383 199 9

Typeset by DP Photosetting, Aylesbury, Bucks

Cover design by Lucy Jenkins

Earthscan Publications Limited is an editorially independent subsidiary of
Kogan Page Limited and publishes in association with the International
Institute for Environment and Development and the World Wide Fund for
Nature.

338.16
F 818 f
35.32

Contents

St. Lucie

3-19-96

1143216

List of Figures and Tables

Foreword

I take great pleasure in introducing a volume that represents an innovative approach to the literature that is building in the area of sustainable agricultural production. David Francis has analysed, on a global basis, the relationship of the social sciences to the ecological and physical problems that have already been well documented. He is a person who has an ideal background and experience for the task, having been trained in both agriculture and rural sociology. His experience includes more than thirty years of grassroots level work in South America, Africa, Central America, and the US.

It was something of a surprise to us and a bonus to our program, to have a visiting professor in the areas of agriculture and sociology from Brazil who was so interested in our local sustainable agriculture research. He brought a new focus to our program in suggesting that the life situations of producers and the economic conditions promoted by government policy must all be examined as part of the same cloth and are critical to reducing the environmental degradation that is of such concern to us all. Family agriculture, while often considered to be old fashioned or inefficient, has great modern potential to resolve problems of sustainability and loss of natural resources. It can also reunite societies around values that have often been somewhat forgotten in the surges of development and the financial debts that have often been incurred in modern farming methods.

I believe that his message has appeal to a very wide audience, both in the US, Europe and worldwide. It is especially timely as many environmental problems have come to the forefront recently and increasing numbers of governments, specialized agencies and concerned individuals are seeking answers to what can be done to bring more satisfaction to our lives even as some are smothered in material goods and others, which are the majority, are finding it more and more difficult to come by the basic needs for survival.

My reading and participation in David's work has increased my own awareness that we are really all part of a common global situation that, although complex and manifested in diverse ways in different parts of the world, is interrelated. All of our actions have their effects on the problem. Even as we suppose that our economic assistance and donations are relieving pressure on the scarce resources of some countries, we create imbalances among those responsible for agricultural production and imply to populations that cheaper resources for food production are available, regardless of their permanence and effects on the environment.

I wholeheartedly recommend *Family Agriculture* to all who are looking for a more complete understanding of world agricultural production and how both producers and consumers have been and will be affected in the future by the way in which we produce our food. I am sure that it will generate broad interest and promote thought on our common future.

Clive A. Edwards
January, 1994

Preface

Many books have been written on the technical aspects of agricultural structure, resource use, economic efficiency and social well-being. This is not one of them. Technical analyses are best for technical people with specified objectives. My goal is to reach leaders who deal with, or have an interest in, aspects of agriculture and rural life as well as those who would like to know more about how agriculture works in various parts of the world. While a good part of the text is based on personal experience, it has been reinforced by the research of some very competent technicians with stimulating ideas.

To make a responsible analysis of agriculture today we must consider the serious problems of decreases in quantities and the quality of resources, increases in population, and the growing differential access that some families are gaining, and losing, from the land. As it becomes apparent that there are complex relations that must be considered, our discussions may also be useful for policy makers and others interested in examining agricultural production from an interdisciplinary perspective. Needless to say, the overview that I propose is not consistent with an in depth analysis of any particular area. But by looking briefly at several issues and areas we can increase our understanding of the total situation and how the parts relate to each other. It is the dynamic change resulting from these evolving relations that will be stressed. Agriculture is being transformed.

The topics that will be examined are focused from the point of view of family agriculture as an organizational structure, not a determined (small) size of production unit. This structure appeared spontaneously and has been used as a component of social systems over the centuries. Each new form of government influenced the character and operations of farm production. The concept of private property, introduced by capitalism in the western countries, has largely determined family agriculture as it is known in these countries today. But the family structure existed long before private property. It was, and is, a structure which produces for the needs of the population and also reproduces the society itself, furnishing workers, duly socialized and with work experience, for the further elaboration of the production sector. And we too are being transformed.

We might then ask: Why is family agriculture under so much pressure today? As new values evolve in societies the existing normative structures are called into question. Some values become less important (for example, freedom from debt); others appear to go out of style (education for its own sake). Structures which perpetuate values that are consistent with the predominant political orientation have the advantage of being promoted. Modernization, economic growth, accumulation of wealth, and other aspects related to what we generally call prosperity have become exceedingly important values in western countries. Family farms attempt to reflect these economic values but, being units involved in both production and reproduction, subsistence and socialization, they find it increasingly difficult to compete in some areas with more industrial forms of agriculture. The result is a sense of alienation as we attempt to adapt to the new orientations.

Current threats to family agriculture in the developed countries give the impression that an archaic structure is headed for extinction. Crises in families also raise questions as to whether a form of production can be based on a partnership that is so often and easily broken. Economic efficiency has increased in importance to the point that free trade is advocated to supply goods and services wherever they can compete with local products – but only during the time that they can compete, and as long as better markets aren't available elsewhere – regardless of the effects that these goods have on other components of local socioeconomic systems. Imported products, for example, may result in a saving in one area only to destroy jobs and the economic environment that is able to sustain communities. Local populations become discouraged and fade away to larger urban regions.

But family agriculture is a global institution that has produced the population, capital, and resources to maintain much of the social and economic development throughout the world. This is not to say that it should be maintained for historical reasons. As we will see as we progress topic by topic through the chapters, many of today's problems existed in previous times and were absorbed by a higher priority, in a structure that was more dear to its participants than the surrounding menaces and distractions.

In much of the world that family structure has been little changed by the encroaching modernization. It is still considered the structure that is most able to produce and most dedicated to protecting national agricultural resources.

The situation, thus, is not hopeless. There are alternatives being experimented with in some countries and traditional solutions used in others that avoid many social and economic crises. There are also some intriguing innovations that can fortify farm families if we are able to seize the opportunities and make them our own. But there is no flow of information concerning successful family agriculture around the world. There is only a trickle of ideas which keep us barely informed that the family is still the predominant structure providing the food and fiber needs of the world's population. Can such an 'old fashioned' structure produce effective results even in the face of enormous population increases as in China, or in areas of resource depletion as in Africa? Can it survive the competition from industrial agriculture with its wealth and power in the developed countries? These are some of the topics that we need to consider.

I invite the reader to participate in the discussion. In that we are investigating topics which vary widely from country to country, and for which change is an essential factor, it would be very interesting to exchange our insights in an ongoing discussion. My US address is: Department of Agricultural Economics and Rural Sociology, 2120 Fyffe Rd, The Ohio State University, Columbus, 43210. In Brazil I can be reached at the: Universidade Federal de Uberlândia, Rua José Ayube, 19, Uberlândia, Minas Gerais, Brazil, 38 400 090.

Acknowledgments

Books represent the collected thoughts of their authors. This book represents the thoughts that the author collected from many others.

I am grateful to my home institution, the *Universidade Federal de Uberlândia*, Fernando Antônio Ferreira, Luiz Antônio Castro Chagas, Adriano Pirtouscheg and Marcos Dias Moreira for providing the opportunity to work on the project. Funding came from the *Coordenação de Aperfeiçoamento de Pessoal de nivel Superior (CAPES)*, of the Brazilian Ministry of Education. The Department of Agricultural Economics and Rural Sociology of The Ohio State University was an excellent host.

Colleagues have shared information over the years, especially in those parts of the world where extended conversations are still popular: Leda de Castro, José Flores Fernandes Filho, José Carlos Araújo, Núbia de Alcântara, Joel Rabb, Larry and Karen Busch, Neal and Jan Flora, Ev Rogers, Dick Meyer, Bob Jacobson, Kamyar Enshayan, Michael Ayi, Zinnah, Adjé, David Kpelle, and many others.

Clive Edwards gave a great deal of time as well as encouragement and direction. Marty Strange is appreciated for his extensive contribution to the manuscript. Terry McCoy and Rattan Lal were encouraging. The US Soil Conservation Service was generous with its publications. John Heyneman brought the related research of Marty Strange to Brazil.

My technical advisors in electronics, Juan and Daniel Francis, made comments on the Information Systems material. Jan Slagle was a friend in deed with her journalistic techniques. My copy editor, Yvonne Clarke, provided very thorough assistance. Lilia and Betsy, the ladies of my life, have been beside me to supply the right words and most politically correct expressions.

I am quite appreciative for all the help and I hope that my interpretations of the information are up to the spirit in which it was given.

1

Introduction

Family agriculture as a form of subsistence is one of the oldest activities known to humanity. It represents a structure that has produced wealth in many countries and formed a basis for the establishment of local, national and even international trade relations. With time the accumulation of wealth was applied to industrial development and manufactured products followed the routes of trade. Then agriculture seemed to become less important. And the family too has changed.

What has been written on the subject is fairly extensive. Much is descriptive, some analytical, some nostalgic and, more than expected, is optimistic. Dr Robert Hildenbrand of Goethe University in Germany wrote: '... in spite of all trends toward industrialization and concentration on a historical and trans-cultural basis, the agricultural family farm has proved itself to be the most satisfying form of rural economy (Hildenbrand, 1989: 159). What does he mean by 'satisfying?' He responds:

> ... there is a marked tendency in West Germany towards the single owner farm. At the same time the number of jobs for farm workers in agriculture who are not owners has decreased from 18 per cent in 1945 to 5 per cent in 1983.... The limits to which agriculture can be industrialized are that in industrial production the machines are stationary and the material to be manufactured mobile while in agricultural production the situation is exactly the opposite: the manufacturing basis, the earth, is stationary and the machines mobile. This is a prohibiting factor in the use of machinery on a large scale.

There has been much speculation as to how agriculture began. The hunter-gatherers of primitive times were nomadic and so would not have developed interests in planting. Fishermen are a possibility – but no society is known to have developed based exclusively on fishing. Agriculture requires discipline that was not characteristic of previous forms of food collection. Seeds have to be planted at the appropriate times, protection and care

are necessary during growth and harvests have to be reaped and
stored. As individuals work together systems must be developed
for the distribution of products. It may be these attributes that
led to subsequent cultural growth. Heiser (1981:15) speculates
that 'It seems quite possible that women deserve the credit,
because they were responsible for the gathering of seeds and
roots and the preparing of meals, and would therefore have had a
much more intimate knowledge of plants than did men'. The real
advantage of agriculture was that it provided more calories per
unit of time and per unit of space than could hunting and
gathering. During thousands of years it was thus considered
natural that small groups work together in a series of tasks that
contributed to their needs for food and fiber and consolidated
what evolved as the family unit as a group capable of survival.
Requirements for production and reproduction, not only physical but social as well, were fulfilled and family agriculture
persisted.

A division of labor evolved in a manner that tasks complemented each other and strengthened the family as a work unit.
'Whose job it is' was a relevant topic for discussion in the family,
with decisions made on the basis of age, gender, physical characteristics and special abilities. These decisions were justified and
legitimized by an evolving tradition. Children from four or five
years of age tended chickens, carried water or hulled palm nuts as
they observed their parents, often involved in food-related tasks
and making decisions as to what area of the forest they would cut
to be burned for the planting of the year's rice, or whether the
back 40 acres would be put in corn again this year. This division
of responsibility served as an example for the elaboration of
production in more complex forms.

With the French and industrial revolutions the family as the
'natural' unit of production came into question. Families fled the
rural areas in search of the promise of freedom in the cities.
Others were forced off the land (for example, in the case of Great
Britain, by the 'enclosures,' the closing of common lands for the
raising of sheep by the aristocratic class). In the cities men,
women and even children were separated from the family to
work long hours in 'sweat shops' to produce manufactured
goods. No longer could they decide how to perform tasks,
neither did they have any claim to what they produced, and
friends and neighbors became competition for the labor positions that promised subsistence and status. In this environment
family agriculture began to be associated with traditionalism,

inefficiency and stagnation. Promoters of new ideas proclaimed that 'you can't stop progress.' In many countries the term 'family agriculture' became confused with subsistence or small farm agriculture. Once identified in this form it was vulnerable to the criticisms of the modernists. The buzz-idea was technology. Technology meant a better, easier and modern life. Technology was applied principally to increase production levels and to reduce needs for labor. The family was thus reduced to one of the factors of production.

And what has all this to do with the *structure* of agricultural production? Structure refers to the way production activities are organized and carried out. Wars have been fought over agricultural structure. The populations of colonized lands all over the world rebelled at paying fees for land use to colonial powers. Slavery as a production structure was internationally condemned and destroyed by armed aggression from external powers, or from within nations themselves, as was the case in the US. Generally the evolution of agrarian structure has not been violent, but it has been tumultuous as families are, for example, forced off the land to seek survival as second class citizens in urban areas.

The production structures put into practice in a society indicate how that society has chosen to meet its needs. Agricultural structure varies in terms of farm size, labor needs and capital requirements depending on what is being produced, whether various crops or livestock are involved (ie the degree of diversification), use of technological innovations, and various other factors. Who owns the land, the labor and the capital also determine structure and vary from nation to nation. Depending on the structure established in a given country, the life patterns of the individuals involved will largely be decided. The possibility of a son staying on the farm, a second son, mutual aid extended in times of need, price guarantees, credit and retirement are all aspects of the structure as it is established at various levels (local, national and international) and times in different places. Societies decide what structure is most appropriate through governmental policies (or the lack of them) and by decisions made by farm families themselves as to the organization of their activities.

Much has been written about labor movement gains – reduced working hours and sometimes the necessity of second jobs – increasing hours again, to maintain a reasonable quality of life. In this process family agriculture was largely forgotten. Various production structures were elaborated as alternatives to meet the needs of an increasingly urban population. The idea of a 'factory in the field' has been established in the production of various crops as phases of the agricultural production process have been industrialized. Ecological problems have accumulated. The urban segment of the population has continued to grow, and is still growing in the developing countries of the world. In the developed countries the family grocery store disappeared along with shoe repairers, mechanics shops and other family controlled firms. Then it was noticed that the family farm persists! In this situation '... theoretical debates about the persistence of the family farm are now emerging from their marginalized status as a "special case", to be hailed as pertinent and instructive...' (Sarah Whatmore, 1991:2).

DEFINITIONS OF FAMILY AGRICULTURE

Considerations by scientific researchers of family agriculture have lacked agreement, initially, in attempts to describe exactly what we're talking about. Some have set statistical limits which determine family agriculture to be 'middle-size' production units versus small farms ('which really aren't farms') and large farms, considered commercial establishments. Other analysts examine income levels based on production sold in commercial markets. Still other researchers have looked at levels of mechanization. Innovative, mechanized establishments are seen in contrast to the 'traditional family agriculture.' A recent definition (Brewster, 1980:19) increased flexibility in the interpretation of the term, declaring that:

> The essential characteristics of a family farm are not to be found in the kind of tenure, or in the size of sales, acreage or capital investment, but in the degree to which productive effort and its reward are vested in the family.... The family farm is a primary agriculture business in which the operator is a risk-taking manager, who with his family does most of the farm work and performs most of the managerial activities.

This is largely in agreement with the concept of the 'commercial family farmer' of Peter Sinclair (1980).

Studies of family agriculture have been categorized into two basic areas of emphasis: 'resiliency' and 'constraints' (Whatmore, 1991). Those who have studied resiliency have often attempted to explain its survival in the face of great social change. Land ownership is considered important as well as the value ascribed by varying structures to family well-being over the realization of profit. Studies based on constraints as a source of transformation of the family structure, on the other hand, have looked more outside the family and see change forced upon it. These researchers study problems of credit, links with agroindustry in purchases as well as sales, and see the family subsumed into a modern system more consistent with urban (profit oriented) industrial production. Neither of these approaches has considered the transformations *within* the family itself and the internal changes that have occurred in the reciprocal processes that the family has with other institutions with which it deals.

The family works out its best way to get the job done. The tasks of each member and relations among members vary within the agricultural families of each nation and generally much more among nations. The American farm wife, for example, becomes a serious bookkeeper to keep up with the paperwork required by the bank. The Spanish farmer's son is attracted to the urban area so the father enters into a cooperative work group with several neighbors to use machinery that he would not be able to purchase for himself and to have additional labor, as needed, from his neighbors.

For this discussion I seek to explore family agriculture on an international basis, defining it in terms of just two aspects: labor use and decision making. While land and water use regulations vary from country to country, farms which are principally operated by family members who are generally free to make decisions concerning production, consumption, storage, commercialization, investments and others of this nature, are considered family agriculture.

GOALS

These aspects and the methods that families adopt for family maintenance, the raising and training of children, the construction of their farming operations, cooperative work groups and other issues of this sort vary greatly from nation to nation. Our goal is to survey family agriculture from a global perspective and discover the similarities.

First I seek to respond to the question, What is family agriculture? Then, Is family agriculture important? Here I will examine social as well as economic aspects of the topic. Finally I will attempt to make some suggestions as to how family agriculture can be strengthened to contribute to a better future for all of us. Although my approach aspires to science, it cannot be considered neutral. It is my belief that family agriculture still merits our serious consideration.

2

Family Agriculture Around the World

While we tend to think of family agriculture as being that which exists in our home country, there are many variations, just as family organization itself varies in different parts of the world. As a concept, family agriculture may involve a midwestern American family on a farm that has been in the family for three or four generations or it may refer to a West African family in an area of common land where a plot of 'high bush' is burned and planted until it loses fertility, at which time another area will be sought. This will not be a systematic discussion of all the types of family agriculture in existence, but a sample based on experience and recent literature in an attempt to examine our first objective: What is family agriculture?

Societies form organizational structures in manners which effectively respond to their needs. The initial requirement for survival in all societies is production. Although family agriculture existed in rudimentary forms before more advanced social organization, it has been nearly universally retained as the form which most efficiently supplies food and fiber, even when sizeable proportions of the population migrate to urban areas. As industry evolved it was perceived that agriculture is inextricably related to success in industrial endeavors due to its influence in the cost of labor. Even today in many developing countries, the majority of the labor class population spends half or more of its income on food and clothing. If the agricultural sector produces inexpensive goods, industry can pay lower wages and compete more effectively in the market. Cheap agricultural products necessitate keeping costs as low as possible. A factor that was early identified as being of importance in the comparatively low costs of family agricultural production is labor. Costs are minimized by using less expensive labor (women and children) for 'chores' and other tasks suited to their abilities, interests and family needs. Costs for training and specialization are borne by the family members themselves as they learn 'on the job'. There is

implicit acceptance of long hours and difficult working conditions. Thus, the family represents an organizational form that
has the possibility of responding to the particular labor needs of
agricultural production, generating inexpensive produce.

Of course food and fiber have not historically been the only
contributions from the agricultural sector to emerging industry.
Many national governments, in need of industrial laborers, have
promoted legislation or have simply allowed a disadvantaged
situation to worsen, in order to encourage an increase in rural-
urban migration. Due to the absence of organization and solidarity in the farm sector, as well as diverging interests and other
aspects to be discussed, farm families have proved unable to
avoid the loss of sons and daughters who enter the urban
environment without the knowledge, skills or experience necessary for survival in the routinized salaried environment and are
relegated to the bottom of the hierarchy.

MODERN TECHNOLOGY

With the technological advances beginning in the 1950s farm
productivity in the developed countries began increasing at rapid
rates, rural-urban migration also increased, and every farmer fed
more people from his production. In this situation the agricultural producer was identified as a prospective client for
industrial products. Research increased and new, specialized
products came on the market. Farm families sensed the necessity
to change long-used methods of cultivation, to tailor the
chemical and mechanical innovations developed for particular
crops or pests, for example, to the crop needs and pest problems
of their lands.

Farm inputs became so specialized that some producers hired
consultants to plan for them. It should be admitted that there
was also an element of status involved. The highly visible innovations especially, were sometimes adopted at levels beyond
what were really the emergent needs of farm operations. The big
blue silos of some countries, bright yellow combines and red
'imported' tractors all represent examples that made neighbors
stand up and take notice – and we all like a little of that. At times
purchases were made in attempts to persuade children to take
more interest in the farm and think seriously about staying on to
continue the tradition. In the process, however, the family was
pushed and pulled by urban-financial-exporting and other
interests under conditions that left it with all the risks that

characterize agricultural production and, at the same time, heavy new responsibilities for production and payments. The results have been problematic. Many have left farming and the relative importance of agriculture has decreased. Pretty and Howes (1993:5) report that:

> *Between 1945 and 1992, the number of regular hired and family workers on farms in England alone fell from 478,000 to 135,000. In the past decade, there have been dramatic falls in the numbers of most classes of people engaged in farming activities throughout Britain.... Fewer people make a living from the land and, of course, they understand it less.*

Most of the new technology was promoted by its manufacturers for its labor-saving qualities. This of course is exactly what was behind its *unsuitability* for use in developing countries, areas with enormous labor reserves and a lack of capital to put them to work. Capital-saving techniques, a more recent concept, took shape only in what came to be called appropriate technology, often springing from the inventiveness of farmers themselves (what can't be done with a piece of baling wire...). With time these ideas were stigmatized, however, as having little potential for increasing production and productivity and relegated to the less developed countries. Looking back over the ensuing situation it might have been wise to have given capital saving technology more attention in the developed countries as well. But the developed country farmer responded to his need for capital by producing more. And as debt increased he produced still more.

With the successes in the developed countries, the farm equipment industry next expanded its market into the developing countries. But the producers who had capital to pay for imported machinery were generally not from family operations. Absentee owners and corporate producers soon learned that machinery developed in the industrial countries worked better with the intensive care and maintenance provided by family farmers. Hired workers didn't fill the bill. From personal observation in Brazil it became all too frequent, through ignorance, negligence or spite, that a rock, for example, could be picked up that would keep a $100,000 + combine out of the field for several days – at the prime time of harvest. Large family operations, however, that purchased and operated modern equipment generally avoided these problems. The dedication necessary for effective maintenance of modern agricultural equipment required an organizational structure in which workers are bound to their activities

by more than a salary. Interest in the preservation and perpetuation of the production process in all of its elements is mini-
mally created by economic incentive alone. Family ties, on the
other hand, have strength, longevity and resiliency to meet the
requirements of modern production techniques. We can then
conclude that another characteristic of family agriculture is that,
because of its internal organization, it is an efficient structure for
the application of modern techniques in the production of food
and fiber.

Societies have manifested agreement with this conclusion in
many ways. The US, for example, paid Revolutionary War
veterans for their time in military service with parcels of frontier
land suitable for farming. This action promoted the idea of the
family as the best organization for agricultural production as
well as settling new areas of the country with productive citizens
who would pledge allegiance to the republic. The country grew
and became stronger.

FAMILY LABOR

In some countries extreme physical conditions of production
have dictated its organization. In his recent book, Harry Oshima
(1987) recognizes the problems of underemployment among
farmers of the 'monsoon agriculture' in Asia. Here the farm
family includes the nuclear family as well as dependent parents
and perhaps grown children who have not accumulated the
resources or found an appropriate situation to establish their
own households. Oshima declares, however, that the conditions
for cultivation of areas with six months of heavy rains and six
months of drought require labor intensive methods. These
methods include planting, transplanting, cleaning and harvesting
rice and other crops almost entirely by hand. He suggests multi-
cropping and diversification with fruits, vegetables and root
crops as well as poultry and other livestock production, fishing
and increases of off-farm employment in labor intensive industry
to make better use of farm labor during the dry season. This is
not to say that he views the requirements of monsoon agriculture
negatively. The intensive work during short periods of time is
seen as promoting cohesiveness and family solidarity.

Attitudes toward cooperation are positive and a high value is
placed on consensus and harmonious relations in Asia. Oshima
contrasts this to the individualism and competition that he
perceives in western, capitalist agriculture. Because of high land

prices people in the monsoon region have small farms and live in close proximity. Their values are maintained and strengthened by the system of Confucian ethics. But when it comes to the importance of the family, Oshima (1987:37) is emphatic: 'Only on small farms with close coordination and the cooperation of highly motivated family members who receive all the returns after paying taxes, rents, and costs, can productivity per hectare rise to high levels in the arduous and demanding husbandry of monsoon agriculture.' Presently the Chinese who practice agriculture in the monsoon regions spin and weave during the dry season. In 1980 the farm population of China was still 70 per cent of the total population.

In nearby Japan family agriculture is carried to the limits of the conceptual production system. The tiny (2–5 acre) plots are the object of intensive labor by all members of the family. There has been some formalized cooperation in terms of classifying, packing and storing fruit and vegetables as well as transport to urban areas, but the farm household remains essentially closed to outsiders as it is felt that their labor intensive activity '... is suitable for each family farm whose members work in close cooperation among themselves' (Ogura and Nakayasu, 1985). Government agencies striving to modernize production are discouraged by the lack of response, partly due to the size of holdings. They pass off resistance with a wave of the hand, hoping that the advanced age of many farmers means that change may become more acceptable in another twenty years or so.

The intensive use of labor is also a characteristic of agriculture in the Amazon region and in many of the traditionally productive areas of Brazil. There, as on the banks of the Tigris and Euphrates Rivers of old, crops are sown on the fertile banks and flood plains when the river recedes at the beginning of the planting season. The fertile layer of soil from up-river eliminates any need to use fertilizer. Weeds and insects have drowned after several months under water. But it is important to plant soon, while the soil is still wet. Intense family labor during this short period soon turns the area green with cash fiber crops, jute and malva, and food crops for consumption and sale in local markets. Land is plentiful so that the family unit is essentially the nuclear family alone. The second period of intense activity comes at the end of the season when the river begins to rise again. Whatever crop remains unharvested may be lost under water during a single night. At that time the pattern of life of the farm family changes. The fibers, actually the bark of the jute and

malva plants, are stripped from the plants and hung in the sun to
dry. The plant stalks are piled up and, if the river rises above
normal, may be used as an island on which to tie the family cattle
and other livestock. A local farmer, when asked what he fed his
cattle while the water was so high, explained that during that
time he goes out in his canoe and collects clumps of green
material that come floating down the river. These plants are
pulled to the livestock for feeding. The farm wife cares for the
smaller livestock, chickens, pigs and goats, to see that they
survive the rising water. The staple foods during this period are
cassava flour in a dry cereal form and fish. These are supple-
mented somewhat during the growing season with greens and
annual food crops. In the municipal region of Manacapuru a
man told me that the river had been exceptionally high that year.
There was a water line on the inside wall of his home nearly thirty
inches above the floor, and the house itself was built on stilts
several feet tall. He gleefully related hooking a three foot fish
from the hammock in the middle of his living room!

The *terra firma* or uplands of the Amazon, although able to
support jungle vegetation, have basically sandy, unproductive
soils. The myth of the Amazon as the breadbasket of the world
was quickly destroyed when attempts were made to establish
modern agriculture in the 60s and 70s. Following deforestation
there is a short period during which the residual fertility can be
used to advantage, then patches of white sand begin to appear.
The family farmers indigenous to the region explain that they use
only the flood plains, which are renewable and renewed. While
mechanization and chemical inputs are virtually unused,
production is at surprisingly high levels – in harmony with the
surroundings. These farmers shouldn't be confused with those
involved in large deforestation projects which are attempting to
bring Amazonas under the plow. They, after a short period of
high income from timber sales, become frustrated with weak and
dying pastures, the brush returns incredibly quickly and conditions, for these pioneers, become unsupportable. The problem is
that too little is known about how to use the enormous areas and
resources of the Amazon to enable current production methods
to operate effectively. Mechanical, chemical and biological
innovations are, to date, no match for the specific characteristics
of the region. Given these circumstances, what type of structural
organization for agricultural production is likely to prevail?

There are examples of the efficiency of family agriculture all
over the world. In Switzerland, Austria and parts of southern

Germany many farms are small and only semi-mechanized. Hay is made where no baler could operate. In valleys where the drying sun is seen for only short periods each day the cut hay is forked onto stakes made from the tops of small pine trees which still have the branch stubs all around to hold the hay loosely for drying. The entire family participates and there is plenty of activity for everyone. If salaried workers had to be hired, no doubt this system of production would be impossible. Hauling the loose hay to the barn adjacent to the house for storage and many other seasonal as well as daily activities are characterized by an intensive use of labor. An old man trimming the roadside with a mowing scythe and putting the cuttings in a big basket to take home to his cattle explained that it helps keep the place clean and, with a wink, 'it helps me out too.'

In many countries the lack of family resources, capital or land results in children staying on with parents even after marriage. This has been referred to in Europe as a 'stem family' (Davis, 1986). The father coordinates activities while the sons work to make savings that can be used to purchase land for their own establishment in farming.

It seems safe to conclude that the family is an effective and evolving structure in forms of agriculture where arduous physical conditions make intensive labor and conscientious management necessary at certain times of the year. The family has also shown that it is able to sustain itself from its labor during periods of slack. With the possibilities of diversification the family has demonstrated flexibility and ingenuity in the development of activities to improve quality of life and/or generate additional income.

COOPERATION

One aspect that has been an essential characteristic of family agriculture historically in many parts of the world is cooperation. In medieval times in Europe cooperation was necessary since there were many common resources in use that had to be regulated by some form of agreement. In England, community owned grazing lands were protected against overuse by specifying the number of animals that each farmer could pasture at various times of the year. In places where solutions to farm problems were beyond the capacity of individual producers, cooperative action built dams, irrigation canals, purchased grain dryers, established retail outlets and conducted many other activities for the common good. In Switzerland irrigation canals are cleaned and maintained by cooperative action in proportion to the quantity of water used by each family. These small farmers also cooperate in sending their cattle and goats with one or two villagers up to the 'high pastures' during the summer months. If every family had to tend its livestock so far from home, much time would be lost and crops in the valley would be neglected.

In France agricultural families have banded together in 'common management groups' to accommodate value changes in the general society where increased importance has been given to individualism, gender relations, and economic expectations (Rambaud, 1985). The role of the State has been important but the idea came from and was put into operation by families themselves since 1965, using their own norms long before the government became involved. By 1983 there were 28,500 groups cultivating over two million hectares and these cooperatives have continued to increase in numbers. Some families enter the group to be able to work off-farm and still keep the farm going. Others have more mechanical and/or labor resources than needed for their land area and can put them to use within the group. Through this cooperative action the family agriculture concept has actually been strengthened. Hired labor dropped from being used on 12 per cent of the nation's farms in 1970, to 9 per cent in 1980. Slightly more than 50 per cent still work full time on their own properties. One characteristic that is given much importance in France is the potential that the groups have for holding families together. While parents are usually separated from their children and brothers and sisters from each other by marriage, the cooperative groups have made it possible for many families to continue family farming relationships. They emphasize that family affection, economic necessity and tradition all contribute to the success of this innovation in the organization of production.

French scientists have noted that the motivational forces used by these farm families are similar to those used to stimulate religious, political and ideological forms of behavior. While all family members participate in management decisions, this is not remunerated. Labor use is negotiated within and among families together with use of machinery and other aspects of the production processes. Women and children especially have expressed increased satisfaction due to their integral participation in ongoing decisions and recognition of the work that they do. Credit, purchases, marketing and machinery and equipment use are all handled as is typical in agricultural cooperatives of other developed countries. Most families report that they have more time for leisure than when they were working individually. The bookkeeping necessary for maintaining fairness among members forces continual, objective evaluation of their activities and recognizes the work of each and every member.

Two problems are noted by Rambaud as being destructive to work group objectives: individualism and divorce. As divorce in France now affects one family in three, the groups must have great flexibility for constant reorganization. Individualistic attitudes have privatized each person as a being 'apart from the church, state, or society' itself. But groups report progress, Rambaud explains, in that their rules change personal family relations into social relations, strengthened by the group as a legitimizing force. The cooperatives have thus transformed the family from within itself to espouse new attitudes and new interdependence, modifying the concept of labor. Although formed to alleviate social and economic problems, these groups have also promoted recognition of the family itself as a viable productive structure.

In Spain cooperativism has existed since the last century. Professor Leovigildo Garrido Egido (1985), agrarian economist in Madrid, explains that longstanding producers of wine and olive oil, for example, have established stable relations of co-operation among themselves, while grain producers, in need of capital for mechanization, have only more recently shown interest in cooperation for completely different reasons. As in France, families began to group together for their mutual advantages before this activity was recognized by the government. In the 1940s legislation created the 'Agrarian Societies of Transformation' through which families working together in the exchange of labor, machinery and land use were eligible for credit and subsidies from the government. Within the groups fields are consolidated although property rights remain individual. Properties in some areas differ greatly in size with some group members contributing proportionally more land while others, with smaller properties, contribute proportionally more labor. Families have shown increased interest in entering the groups when off-farm opportunities for part-time labor are more abundant. Initially the groups generally do not provide full-time employment for all members. Decisions are not made 'democratically' on the basis of one man-one vote as might be expected in cooperative groups, but depend upon the capital and land

values of each member, as are the criteria for the distribution of
returns. Some groups provide workers with a share of returns,
others pay only hourly wages.

Professor Garrido Egido (1985:42) declares that voluntarism is
fundamental for stability and longevity of the group activity. It is
important that there is some family, neighboring or friendship tie
among group members since 'Human relations are the most
important factor in the functioning of a group and sometimes
when the leader disappears, who was its creative and motive
force, the group begins to have problems and drifts toward
dissolution.'

With entrance into a cooperative group some families stop
cultivating land to concentrate on livestock production. Thirty
years after the agrarian societies were established cattle
production had increased by 3.5 times; sheep by 2.5 times and
swine by 15 times. Irrigation systems were also constructed which
required additional labor. With time, participants age and begin
to hire labor – this is necessary because the sons of group families
often do not work in the same groups with their families. Garrido
Egido also admits that there has been some corruption; groups
have obtained credit and subsidies without really working as a
unit. In general, however, government objectives have been
reached as agricultural production has increased and been
mechanized.

In Denmark, according to W P Watkins of the International
Cooperative Movement, cooperatives have attempted to modernize with the times (Watkins:1986). The spirit of cooperation is
strong in the society in general and especially among families,
who average 30 hectare (70 acre) operations. Technical communications from the cooperative assist decision making for
planting and livestock raising. Collective decisions may be made
to limit production and avoid saturating the market or to enter
forcefully in a production area to take advantage of market
opportunities. Agricultural innovations and new ideas for
farming are promoted by the cooperative itself. In the case of
equipment purchases the cooperative generally negotiates for all
members, as a unit, to obtain lower prices. If credit is necessary
that too is handled within the coop structure. Members thus
plan, borrow, loan, buy and sell to each other as well as to the
outside market, all through collective decisions made by members in similar situations with similar needs, objectives and
interests. This cooperative structure as a voluntary, flexible
institution has become solid tradition through which the society

assures protection to rural producers and autonomy for family agriculture.

Unfortunately the formal cooperatives in some countries of the developed world have not been able to maintain bonds of loyalty. They have been weakened by their lack of ability to attract great numbers of farmers, just as the farmers themselves have been weakened by financial and economic problems. Through all this it is easy to say that solutions lie in increased organization and cooperation among all farmers, but it is, to be sure, very difficult to realize change.

In the developing countries a notable characteristic of the most successful cooperatives is that there is some special unifying attribute. Ethnic background, where diversity exists; specialized production needs; or particular concerns that set members apart from the general population strengthen unity for the realization of effective cooperation. Japanese families in São Paulo, Brazil, and agricultural extension workers in Togo, West Africa, are two examples from my own experience. This appears also to have been the case, historically, in the Yonland of the United States, as reported by Kraenzel (1980). The Hutterite, Mennonite, Mormon and Amish groups effectively united in cooperative action. But the extension of this activity to those outside did not occur.

Thus families, even as they have been transformed by changes in their surroundings, have demonstrated the capacity to come together to increase efficiency and levels of production through cooperatives.

Informal Cooperative Efforts

Cooperation springs from necessity, and it has also developed to reduce the burden of work and increase camaraderie. In West Africa cattle and horses are rare because of disease problems, consequently all hauling and farm work are done by hand. Since the topsoil is very shallow, preparation for planting crops really consists of scratching the freshly burned soil surface. The planting of a 3–5 acre family plot would be considered therefore an overwhelming and solitary task for one family. Families work together in groups of twenty or more, mostly composed of women, who work side by side spreading the ashes in the wooded areas that the men have cut and burned during the dry season, and fluffing up the soil for planting as they sing and tell stories. There are various aspects of cooperative action that become

apparent in this situation. The individual producer, while getting
the area ready for burning, is aware that many friends and
neighbors from the village will soon be there to plant and they
will comment on the size of the area and how well it has been
prepared. He is thus highly motivated to do a good job, not only
for the subsistence of his family, but to enhance his status in the
village community as well. Sociologists refer to this type of
activity as being influenced by social effects: individual behavior
is modified by the sentiments and possible impressions that could
be created.

Families also come to know each other very well in terms of
work habits and capacities as well as singing and storytelling
abilities. In this situation the community of families is bonded
together in a union that is not without conflict, but is sufficiently
strong to meet the conditions of nature. They share very pleasant
times and they share misery and disaster as well. Hunger is an
annual occurrence since harvests are meager and grain storage is
difficult. In a village where I spent a year setting up a regional
school and local agricultural extension service a woman gave an
example. A group of workers left the village one morning at the
beginning of the wet season to prepare a family's area for
planting. The area was more than a kilometer away so it was
necessary to get an early start. They walked happily, laughing
and joking with one another. One young mother was the object
of much conversation. She had given birth only two months
before and there was doubt that she would be able to keep up
with the others. The man who had burned the plot led the group.
Several carried drums, iron bells or glass bottles to tap to provide
a rhythm for the work. The area had burned well. Pieces of
trunks had been pulled to the remaining stumps to finish burning
and leave the area cleaner for planting. It had been a lush area so
there were piles of ashes to be scattered, providing fertilizer for
the rice crop. Arriving at the scene the women put their
belongings, some palm wine to restore energy and the baby in the
shade of a tree on the edge of the clearing. As they worked the
baby was heard crying for a while but then apparently dropped
off to sleep. When they stopped for a rest they noticed driver ants
under the tree. These are the ants that can strip the flesh from
even large animals in a short time. (It is said that after the big boa
constrictors make a kill, they circle the area searching for the ants
before consuming their prey. After eating they become very
lethargic for several weeks and would be vulnerable to an attack.)
The mother ran to the tree where she saw to her horror that the

ants had indeed attacked the baby. There was little left except the cloths in which he had been wrapped. The others rushed to her aid. Although her grief was enormous, it became bearable as it was shared throughout the group. They returned to the village crying, screaming and moaning. The women shaved their heads and went into mourning. While this example is very different from the kinds of problems confronted by families in other nations, it demonstrates that family agriculture can be strengthened by families working together in cooperative effort.

Groups of farmers also unite for cooperative activities in Latin America. In Brazil this group is called a *mutirão*. It is temporary and formed at times when a lot of work is needed during a short time, such as planting and harvesting. As in Africa, there is a certain social influence as farmers perform their tasks knowing that their colleagues will be observing what they have done.

These activities also existed in the US during the time of the threshing rings. Families united to thresh their grain using the mechanical innovation that they often purchased jointly. The events were celebrated at the time and are still fondly remembered in the grandparents' photo albums. Such group activities furnish an excellent example of social effects as not only the diligence and stamina of the participants were observed but evaluations were made. There was talk of the 'real go-getter' who 'pitched in' and 'took the bull by the horns' to 'make the dust fly,' who didn't 'put all of her eggs in the same basket,' or 'cry over spilled milk,' who could 'get in harness,' 'put his shoulder to the wheel,' to 'make hay while the sun shines . . .' that is if he or she could 'buckle down' and 'get on the beam.' Many of these expressions remain with us even today. One aspect of the threshing rings that enhanced social significance, as in West Africa, was that women participated. The midday meal was a notable occasion; it was a time to get away from the work for recreation. Dishes that were appreciated were roundly praised while just one person holding up a pin feather and asking 'Who picked this chicken?' could cause a ripple of embarrassment.

As can be seen from this discussion, group action doesn't necessarily mean cooperatives. Institutionalized cooperatives formalize action to realize mutual benefits. They have been of great service to family agriculture in the developed countries and should play an important role in the future. But as cooperatives have grown and become stronger the face to face contact among members has been reduced. The effects generated by personal interaction among families in informal cooperative work groups, on the other hand, have a special influence. They bind the participants together in meaningful association. Individuals do things that they would not have done had they not been together. With the degeneration of face to face contact other forms of interaction also spiral down. Visiting patterns are weakened. Exchanges of presents and concern among families become less frequent. These effects coupled with improved possibilities for transportation and communication have ironically resulted in the general isolation of farm families from each other in the developed countries. The effect has been to weaken the participation given to institutional cooperatives as well.

Informal cooperation has existed among Norwegian families for generations. Dairy farms consist of approximately 20 acres and 10 cows, so families live fairly close together. With the implantation and increasing importance of market relations, however, Dr Reider Almås at the University of Trondheim reports (1985) that mutual assistance in the form of labor exchange and threshing ring types of work have been reduced. Among producers of irrigated crops, on the other hand, the construction and preservation of the canals has been a source of longstanding cooperation. Dr Almås sees the history of informal cooperation with its family and neighborhood ties as having been important for people to come to know and work with one another 'as stable and reliable partners.' This experience sometimes leads to the establishment of formal cooperation. Today formal cooperatives have become popular especially on small and middle-sized farms. Farmers show great enthusiasm during the organizational phase, then, with the routinization of activities, participation begins to fall off. There is, however, a rallying in any time of crisis. If this evokes a sense of 'just like here,' it is because cooperation as an institutional form has become a victim of more individualized forms of production in many countries of the world.

GOVERNMENT-INDUCED STRUCTURAL CHANGE

In some countries 'cooperatives' have been established by governments for their own objectives and convenience. Several of the French colonial countries in Africa established what were called agricultural cooperatives, meaning essentially that producers cooperated with the government, receiving access to modern techniques, seed and fertilizer. In the Amazon region of Brazil 'cooperatives' established by the agricultural extension

service received presses to remove the moisture from fiber crops –
jute and malva – in preparation for shipping to market. But when
an attempt was made to interview members some years ago none
could be located. The usual response was that the coop treasurer
had disappeared with the funds and the press had been abandoned. In Egypt, cotton and rice producers receive credit,
improved seed, pesticides and fertilizer from cooperatives and
are required, by law, to sell their produce back to the cooperative. The government, in turn, claims a percentage of the product
as a form of taxation, a more efficient system than that which
attempts to tax individual producers (Radwan and Lee, 1986).

These examples, however, demonstrate little of what cooperatives are really about. A brief look at cooperative principles
provides an explanation. Cooperatives, as a means of promoting
family agriculture, must be organized from the grass roots level.
Only when those who are affected by the decisions (costs and
benefits) are integrally involved in the ongoing decision making
and operational processes, are the real principles of cooperation
realized.

In China, the decade of the 80s brought governmental
relaxation of the collectivization of family agriculture in an effort
to increase production. Families have responded positively to the
possibility of independence and production levels have been
raised. The World Resources Institute (1992) reports that the
Chinese per capita food production (in spite of a population of
more than one billion) has increased by 50 per cent during the
last twenty years. Now, in the 90s, the tendency is for the ex-
Soviet bloc countries, pressured by the same urgent necessity for
more production, to do the same. While there have been varying
degrees of success with collective farms among the Soviet
countries the need for increased production makes some policy
change imperative.

An interesting example of family agriculture in the ex-Soviet
bloc is to be found in Hungary. This country has had considerable success with agricultural production on its collectivized
farms. Domestic consumption is plentiful and accessible. The
country has a positive and sizeable trade balance both with the
industrialized countries to its west and with the recognized ex-
Soviet countries to its east. Agriculture provides more foreign
currency earnings than any other sector of the economy. Nearly
half of the Hungarian population is associated with agricultural
production either directly or indirectly. In this situation Dr Antal
Gyenes, director of the Cooperative Research Institute, contends

that the individual producer, working with members of his family
on their small, household farms, is still the most industrious and
most efficient method of production for saving energy and
material (Gyenes, 1985). But little chance is seen for Hungarian
farms to revert from their socialist experiences to the system
prevalent before Soviet occupation (Gyenes, 1985:236):

> The Cooperative Research Institute has been conducting a
> survey every fifth year since 1967 in which interviews have
> been conducted on the basis of the same model containing the
> question intended for the former small-scale individual pro-
> ducers and their children: Would they be prepared to restore
> private production? Way back in 1967, six years after the
> completion of collectivization or transformation of agri-
> culture, those who wanted to return to private production out
> of the members of the farmers' cooperatives interviewed
> accounted for just over 2 per cent. Since then their proportion
> has been reduced to zero according to repeated interviews.
> Thus it would be impossible to turn the clock back.... Pea-
> sant traditions have undergone profound changes, for instead
> of being motivated by the desire to accumulate resources
> which are indispensable to acquire land, a characteristic effort
> looking back on century old traditions, the peasants have
> turned towards consumption. They have built new houses and
> have made every effort to make them as comfortable and up to
> the mark as possible. However, the prerequisite of this was to
> perform an increasing amount of work in the large-scale
> farms besides working as much as they can in their own
> household farms.

Szelényi (1988:4) explains the political side of the Hungarian
situation. He sees a 'Third Road,' a mixed economy (capitalist
and socialist), as offering a more balanced distribution of power
and more opportunity for those at the lowest socioeconomic
levels:

> The dual existence these peasant-workers have established for
> themselves has many advantages: state employment guaran-
> tees a permanent flow of cash income into the household and
> insures against the uncertainties of the climate and the
> market. Why should they give it up? Furthermore, it is a long
> way from their highly intensive, productive and profitable, but
> tiny family minifarms to a viable full-time family farm.
> Without a proper commercial credit system and with legal
> restraints on land ownership, our new entrepreneurs would
> find their opportunities to become full-time family farmers
> very restricted indeed, even if they wanted to.

Recently excess farm labor has found new opportunities for remuneration in industry. But families retain their rural base and reduce their part-time urban employment when needed for agricultural activities. Local urban planners note differences from other countries in that industry has developed without the accompanying urbanization that usually requires high levels of investment and alterations in the traditional rural society. While it is true that no country is purely capitalist or socialist (even in the United States there are few who would advocate elimination of social security, a typically socialist measure), Hungary has evolved a system which, for family agriculture, combines the advantages of both capitalism and socialism. The families, for their part, have been found to represent a form of organization that can be integrated into varying political structures as a component for the realization of agricultural production.

FARMER-INDUCED CHANGE

It is through the union of forces and face to face contact that exchanges of new farm ideas occur. The diffusion and adoption of innovation studies conducted among US midwestern farmers since the 1950s have established that the great majority of farmers get new ideas from family members, neighbors and friends (Rogers, 1983). This is true around the world. People with similar histories and interests are perceived as being more credible. Sharing new ideas strengthens relations and promotes the perpetuation of the structure, in this case family agriculture.

While results of this sharing of ideas are nearly always positive, how the ideas will be integrated into family operations is sometimes unpredictable. The critical need for protein in West Africa provides an example. Emaciated children with hair that loses its blackness to a dull red indicates protein deficiency. Peace Corps volunteer Burt Green's project to alleviate this problem and contribute to community wellbeing some years ago resulted in the construction of a 'laying hen' project at the village school in Ganta, Liberia. Hens were donated by the Israeli Agency for International Aid and a balanced diet was formulated using cassava leaves and local grains. Eggs were sold for very low prices in the village market. Burt was sure that as people ate more eggs their health would improve. But when first evaluations of the acceptance of the eggs were conducted it was discovered that the people were not consuming them. Their purchases were being put under little native hens for incubation because 'We want big hens like those at the school.' The problem was that the project had not foreseen this possibility and had brought no roosters from the Israelis. Attempts to convince the people that these eggs were not fertilized and for consumption only were passed off: 'You foreigners know about many things, but here we know how to raise chickens.' A month later there was a faint smell of rotten eggs in town and no one wanted anything to do with the eggs from the school. The project was discredited. By the time roosters could be brought in, eggs were piling up at the school. Subsequent attempts to assure people that the eggs could now be put under their hens for hatching were of no avail. Finally donations of several dozen eggs were made, first to the village chief and his wives and then to the local military authorities. Twenty-one days later bright yellow chicks, twice as big as the native chicks, were running along with local hens and project interest was revived. The objective of applying innovative ideas for increasing protein consumption was actually attained, and chicks from the school hens were observed in villages as far as forty miles from the school. But the families involved revised Peace Corps methods to serve their own needs and interests.

Sometimes the spread of new ideas and practices can present problems. In Brazil there are land reform projects based on the concept of the 'agrovila.' These towns of 250 families house the people together for service delivery, as is their custom. Each family, recruited from landless segments of the population all over the country, receives a house and a plot of 20 hectares (50 acres) in the rural area adjoining the town. At the Serra do Ramalho project in the State of Bahia one of the agrovilas is close to a large commercial cotton project where central pivot systems irrigate crops for production that go directly to textile industries in the southern part of the country. The family farmers of the agrovilas have also discovered that cotton is a profitable crop. There are no weevil problems as exist in other parts of the country and marketing channels have been established. In that situation one can speculate as to how long it will be before a family, with the goals of improvement and increased production, inadvertently brings in contaminated seed or plant products from other parts of the country and introduces the weevil to the cotton in the region.

In southern Brazil and Argentina a recent crop innovation has been the cultivation of soybeans. Studies by the Brazilian Institute for Agricultural Research show that the climate may be the best in the world for soybean production. While soil conditions require some modifications, lime deposits are found in the areas so that acidity does not present a serious problem. Under these conditions many traditional ranchers have plowed up pastures, or rented them to be plowed. Soybean production and exports increased greatly in the 1980s and there was general optimism. The price subsequently dropped, however, and producers reconsidered their decisions. Today many raise soybeans as a profitable legume to fix nitrogen and fortify soils for better pasture. They say 'the beans pay the costs' of improving pastures. Again agriculturalists have adapted the new ideas, promoted by those outside of agriculture, to their own traditional needs and interests, and will continue raising cattle.

These examples strengthen belief that family agriculture is a valuable structure for the evaluation, adaptation and modification of agricultural innovations to improve methods of production and the well-being of the farm population as well as the population in general.

RESPONSES TO CHANGE IN SOCIETY

The importance given to family agriculture in Europe is represented by recent happenings in Sweden. David Vail (1986) reports that loss of farmland 'either to forest or to asphalt' has been a national concern of high priority since the 1960s. He explains that in 1979 a law was passed to empower county agricultural boards to purchase farms on the market to avoid sale to non-farm investors, and offer them with favorable mortgage rates to those aspiring to enter farming. The law was thus established to 'limit land speculation and reaffirm commitment to family farming.' Farming methods have improved and the country has largely attained self sufficiency in foodstuffs, with costs dropping from one-third to one-fifth of consumers' disposable incomes. A negative aspect, however, is that a 'grain glut' has accumulated. A serious problem in Sweden, as in most of the developed world, is overproduction. Protection of the well-being of the farm population has become increasingly difficult as it diminishes to a small fraction of the total population. Negative reaction from the urban sector to traditional farm policies, such as import quotas and price supports, becomes inevitable and conflict builds. It is thus understandable, although quite surprising, that the leaders of the Swedish National Farmers' Federation took the initiative to propose the planting of trees on cropland. Reforestation was proposed to be used on selected lands along with other techniques including the production of high protein fodders as a means of partially substituting imported protein (soybeans), and agro-bioenergy crops as a contribution to national self sufficiency in energy and, at the same time, to keep agricultural lands in use.

Acceptance by farmers of 'reduced input' production – less soil preparation, use of fertilizer and agro chemicals – has promoted a union of forces among those with interest in farming, ecology, and the protection and revitalization of rural communities.

Sweden is heavily dependent on nuclear sources for energy. While there is increasing pressure to reduce this dependence, there must be feasible alternatives. The Soviet nuclear accident at nearby Chernobyl contaminated large areas of Swedish agricultural lands and called attention to the importance of reliable national sources of food and energy. Alternatives are also being sought to maintain the farm population in place. Integration with other sources of employment – forestry, tourism and crafts – is being considered. The present situation is seen by some, Vail concludes, as the beginning of new opportunities to consolidate political coalitions and develop strategies that were previously not available for the realization of a rural sector in keeping with evolving national values and objectives.

Historically there have been many cases in which family agriculture has appeared to fill a void in processes of development. In Latin America, for example, the early mining, logging or even export agriculture projects seldom planned for the local production of food – meat as well as grain, fruit and vegetables – for workers. Early history books relate that food necessities were either imported from Europe or brought in from more agriculturally developed regions. It was through the efforts of families in the area that lands were cleared and crops and livestock raised for sale to the local colonial projects.

Sugar production in Taiwan, Republic of China, represents another example in which plantation type cane production was not feasible due to limits in the quantity of available land. At the same time it was economically not practical for small producers to purchase equipment to extract and refine the sugar. An agro-industrial system has evolved as a solution: cane is raised by families who negotiate for the sale of their crop with the sugar refineries (Ka, 1991).

These examples all demonstrate ways through which family agriculture has diversified to complement other production structures, with the result of increased functionality in the total production system.

CONSERVATION OF RESOURCES

A final characteristic of family agriculture around the world is its ability to conserve resources. The passing of lands through the generations, each hoping for a better future for children and on through the grandchildren, creates a dedication that is universal yet unprecedented in other production structures, to use only what is necessary and maximize resources for the future. This was very apparent from living two years among the Cabrai people of Togo, West Africa.

Located close to the southern edge of the Sahara Desert, these family agriculturalists construct stone terraces to secure the limited soil that washes down the hillsides. Terraces are repaired regularly and, in spite of the absence of formalized private property, these lands remain in the same families for generations. If a gully starts to form it will be blocked into a small triangle and subsequent parallel barriers built below it to a terrace of two feet in height, depending on the slope and the quantity of stones in the area. Some terraces may hold only enough soil to raise one plant. The objective is to conserve all soil for production. Every family has one or more stone lined pits close to the house. These pits, six to eight feet deep and six feet across, are used for the accumulation of compost. Beginning when the sorghum (a staple crop that is consumed like the cornmeal 'muah' of other coun tries and also used to make beer) sprouts in the fields, goats are tied in the bottoms of these pits to keep them out of the fields and are fed the lower sorghum leaves. The plants grow to a height of eight to ten feet. Other household refuse (ashes, chicken feathers, peanut plants and hulls etc) also go into the pits and the goats grow and slowly emerge to the ground level as the pit fills in under them. After the sorghum harvest the pits are nearly full, there are no more leaves to feed the goats and they are turned loose. Through the dry season the pits are covered with plant trash to protect the contents for the next planting season when the decayed material will be taken out in large baskets and applied around the seeds and cuttings of the new crops. The children participate in this process, learning the production and conservation practices of their parents and knowing that much of the work is for their future benefit.

Over the years the differential soil qualities between the sandy paths and the crop soil in the narrow terraces is amazing. The present generation owes much to its 'ancestors' and honors them by spilling a little of the sorghum beer ('*tchoucatou*') from the calabash in which it is drunk. This respect is shown every time the *tchoucatou* is consumed, which tends to be fairly often!

The sorghum stalks are cut and lodged in bundles among the branches of the huge baobob trees to be used as needed for firewood. Nothing is wasted as in this barren land sources of energy are severely limited. It is not difficult to imagine, then, the local sentiment when foreign aid technicians attempted to introduce the cultivation of a very productive variety of dwarf grain sorghum. Dwarf plants have few leaves that would feed few goats and make less compost. There would be no firewood, but more grain – which is very difficult to store in a way that is protected from insects – and, in any case, cannot be eaten raw. Better for the grain borers, a disaster for the people.

The response by government planners and foreign advisors, when advised of these details, was that production of grain could be increased for sale in the export market making it possible for the families to purchase their needs. But this was not acceptable to the Cabrai families. Their modest system works, and will continue to be used.

Conservation is most practised in places where resources are scarce: in the Appalachian region of the United States, in the rocky fields of Haiti, even in the small Imperial Valley of the ancient Incas in Peru, where the mountain streams come together to form the source of the Amazon River. Resources are protected for production now and for the future. Without doubt, family agriculture is unprecedented in its ability to conserve plant, animal, land and water resources as well as to perpetuate an evolving knowledge of production methods and techniques of conservation.

SUMMARY

- The family represents an organizational form that has the possibility of responding to the particular labor needs of agricultural production, generating inexpensive produce.
- Because of its internal organization, the family is an efficient structure for the application of modern techniques in the production of food and fiber.
- The family is an effective and evolving structure in forms of agriculture where arduous physical conditions make intensive

labor and conscientious management necessary at certain times of the year. The family has also shown that it is able to sustain itself from its labor during periods of slack.

- Families, even as they have been transformed by changes in their surroundings, have demonstrated the capacity to come together to increase efficiency and levels of production through cooperatives.
- The effects generated by personal interaction among families in informal cooperative work groups have a special influence. They bind the participants together in meaningful association.
- Families have been found to represent a form of organization that can be integrated into varying political structures as a component for the realization of agricultural production.
- Family agriculture is a valuable structure for the evaluation, adaptation and modification of agricultural innovations to improve methods of production and the well-being of the farm population as well as the population in general.
- Family agriculture has diversified to complement other production structures, with the result of increased functionality in the total production system.
- Family agriculture is unprecedented in its ability to conserve plant, animal, land and water resources as well as to perpetuate an evolving knowledge of production methods and techniques of conservation.

These are just a few of the numerous characteristics that agricultural families around the world demonstrate in the production processes of their daily lives. Others will be discussed in the next chapters. The point is that the family is not just an organizational structure that is disappearing from agricultural production in the developed countries. It has existed historically in nearly all parts of the world and continues, not only producing the resources that have made economic development possible, but also as a social structure that accumulated customs which became the prototypes for national values, as surplus production made urbanization possible. It is this social role that family agriculture has performed that will be examined in the next chapter.

3

The Social Role of Family Agriculture

In the Introduction I stated several goals in the form of questions for our discussion, among which was the importance of family agriculture as a structure in society. This chapter will begin to respond to that question.

The fact that family agriculture has a special role to play in society is associated with its historic presence as a productive unit capable of devising systems for labor use as well as product distribution. These systems serve as models for other organizational forms in society. In the processes of agricultural production the social and economic roles are inextricably intertwined. Many researchers, however, admit to the importance of only about half of this relationship. For them and those who follow this line of thought, production and efficiency suffice as goals for the agricultural sector. But in countries where subsistence agriculture is still prominent there is difficulty in understanding the need for economic efficiency. This is a concept that becomes more important when the product enters the market place. Families in developed countries also make decisions without regard for efficiency, keeping a few laying hens, or a pony ... just because they are valued above principles of economics. Organic producers, as judged by these economic criteria, might not even qualify for a farm loan. One way to understand the relation of the economic to the social roles in agriculture is to see the economists as dealing with the *means* for realizing the material needs of society while the social role pertains to the *ends* or objectives for which we strive and which have served as models for each society.

In recent times various areas of the social sciences have become interested in agriculture. Even philosophers have become involved. One of them is Paul Thompson, Professor at Texas A&M University. Thompson (1986:41) believes that:

> *These economic goals are* real *goals to be sure.... But we will never find our complete salvation merely in the right set of*

economic policies. To make an agriculture that will serve our
need for a spirit of community and self-reliance in the future,
we must first accept the need *for community and self-reliance*
once again as social and moral goals for agriculture.

Thus for Thompson production and efficiency (or profit at the
individual level) are only one part of what agricultural production is really all about.

Cornelia Flora, a sociologist in Virginia, writes on the
importance of the various activities of farmers. She broadens our
understanding by explaining that its not just corn and soybeans
that are produced. The purchases of farm equipment, seed,
fertilizer, bank credit, sale of products to grain elevators and
participation in social activities all contribute to the production
of community and form a base for meaningful social interaction
(1986).

The emphasis given by both Thompson and Flora to community vitality should not suggest that it is an important social
component of family agriculture alone. The strengthening of
community relations integrates all populations to reinforce social
values. Individualism is kept under control and there is conscious
effort to maintain positive social contact with neighbors and
friends. Traditionally family agriculture has set an excellent
example for the greater society in this area. We stress the
importance of community here because farmers still need these
relationships. And society, more than ever, is in need of positive
guides.

AN HISTORICAL REVIEW

From this point of view, what has been the historical role of
family agriculture? First, as we have seen, the family provided an
organizational structure for the realization of production, the
initial system-problem of every society. Although the family
functioned as a production unit well before feudal times, an
interesting example is cited by Davis (1986) of the feudal lord of
Duino in northern Italy, close to the seaport at Trieste. Serfs
were charged with furnishing all needs of the castle and chapel
including olive oil, firewood, wheat, rye, vegetables, money and
even the carrying of water. For these tasks it was imperative that
the families not only produce but reproduce as well, to guarantee
a supply of servants for the lord. To this end he summoned

young unmarried men and women serfs to the castle once a year, paired them off, and ordered them to marry.

The family also created a nucleus for elaboration and perpetuation of the social norms that provide stability and continuity in society. Children observed and learned from parents and repeated the processes of cultivation and livestock care as well as the social rituals and, in the above example, manorial requirements. In time these simple norms became formalized and ritualized into culture. Culture, in turn, acts upon the family as the society becomes more complex. Farmers are motivated, for example, to increase productivity to win a prize in the local corn producers' club. Making money and profit may become goals of importance as never before.

There are also goals for more than production and profit. In modern times consumption has become an increasingly important aspect that must be organized. The family is a consumption unit *par excellence*. As extended families have been reduced to nuclear units and are getting closer to Toffler's 'one person family,' the number of families has increased and every unit aspires to obtain all of the basic goods and services of the market. Farm families are especially adept as consumption units because the sum of the workplace to the household increases need. The concept of need has actually changed as needs are 'created' together with the modern products. Today we have needs that had never been heard of twenty years ago, and these needs increase the importance of cash income and thus the deprivation that family agriculture senses when material living standards are compared with urban salaried workers.

Dividing Up the Tasks

The agricultural family has also been important to society as a role model for the division of labor. Tasks were first divided between genders and ages in accord with the human resources available. While this is common knowledge and practised globally in family agriculture, the Guaymi Indians of Panama present a simple and convincing illustration of the significance of a division of work. Reverend Ephraim Alphonse who has worked among the Guaymis for more than sixty years explains that during the dry season, before planting time, the men choose plots of land that appear to be fertile and 'clean' the areas for cultivation. The land belongs to the tribe and is used as needed, by collective decision. If land used the previous year produced well, it will be planted again. It may not be necessary for the family who worked there to clear additional land. This doesn't involve cutting and burning such large trees as is true in Africa, explained in the last chapter, but the principle is the same – bush is cleared away and the area prepared for cultivation.

Women then take over soil preparation and planting activities while the men divide their time between home activities, chipping out dug-out canoes, doing any building and maintenance that the family unit may require, and fishing in the ocean. The weather and conditions of the sea determine the daily occupation. From their activities the women bring yams, green bananas or cassava 'to fill the belly' as well as greens, vegetables, fruits and pepper from the farm. The men bring the fish. 'If this sounds to you like women work while men play,' Reverend Alphonse exclaims, 'you have never gone to sea in a dug-out canoe!' The diet changes during the year as vegetables mature, fruits ripen and different species of fish are available. There are happy surprises as well as crises as the work of each of the family members complements that of the others in the division of labor that has survived for generations.

Knowledge of the family division of work is critical for extension workers and other change agents in contact with agricultural families. A regional program for the control of brucellosis in the dairy cattle of southern Brazil completed two years without making substantial reductions in the disease. Government extension agents reported in a national conference that they gathered one day to debate the future phases of the program. They were disappointed since the men who had attended their meetings were quite open and appeared to accept the practices necessary for disease control. As discussions unfolded, someone remembered it's the *women* who milk and care for the cattle. Although some women had attended the extension meetings there had been no effort to invite them or orient the program to their special needs. With revision the program results improved.

Thus the family, through its agricultural pursuits, has developed a division of labor for the realization of production and perpetuated this system in the family structure through reproduction and socialization of new workers. It is these evolving normative systems that have been, with time, formalized into culture. As other organizational forms of production have evolved they have been built upon that simple and effective structure developed in family agriculture.

Another expression of this historical role has been the construction of a base for social stability. With stability, of course, other possibilities are created for the elaboration of complex society.

The benefits of family agriculture as a model for social organization do not exist only in history. To be more specific, in a present day issue such as gender relations, how does family agriculture provide influential examples in today's societies? The farm wife, first of all, is involved in the planning of family activities. This varies considerably from culture to culture. The rural West African woman, for example, works together with her husband on the quantity of land that they are able to cultivate and hopefully accumulate a surplus. At that time it is often the wife who initiates discussion concerning a single neighbor or a friend's daughter who might fit into the family to reduce the level of household work, take care of the baby, and make it possible to cultivate more land. With two wives the surplus may increase, making possible additional increments to the family unit. In every instance the husband and wife or wives plan together. If circumstances permit the family to grow, the first or head wife attains additional responsibility and status. She handles the money for children's school uniforms, books and supplies as well as for household expenses. This is not to say that other wives do not participate in the family decisions, but this will depend, however, to varying extents on the head wife. In some families she develops special skills to promote and protect her family, for example the application of patent medicines for minor health problems, and the regular distribution of anti-malaria pills. She will also mediate any misunderstandings between her husband and other wives or family members much as a marriage and family counselor would in other societies.

While farm planning is in some ways less complex in more developed societies, the farm wife is often the family business manager and budget officer. This 'helpmate' relationship constitutes a base for mutual support and mutual respect. Each of these examples illustrates how the family structure furnishes a model of how individuals can work together for their mutual benefit. Responsibilities differ, are more or less democratic, follow, and mold new social orientations for the larger society.

Self Reliance and Other Virtues

Thompson also cited the need for self-reliance as a social goal. It is the physical situation of family agriculture that builds reliance on one's self. In contrast to salaried work, there is no boss nor directives to be followed. Jobs well done reinforce the individual's self image and that of family members to do more, just as too little millet saved for the dry season or water left turned on in the calf pen can't be blamed on anyone else.

Thomas Jefferson mentioned the positive character traits found among family agriculturalists in a letter to his friend, John Jay, in 1785: 'Cultivators of the earth are the most vigorous, the most independent, the most virtuous, and they are tied to their country and wedded to its liberty & interests by the most lasting bonds.' Thompson (1986:37) explains that

> *Like many moral and political theorists of his time, Jefferson was mindful of the importance of self-interest in individual decisions. He and the other founding fathers saw their task as one which would marry self-interest to social unity (and, thereby, to a broader concept of the good).... Since a farmer must stay in one spot, he must learn to get along with his neighbors and take an interest in long term stability. The virtues of honesty, integrity, and charity promote a stable society, and are also the virtues that promote the farmer's own interest.*

While the historical role of family agriculture remains valid, it is not enough to examine the past and use it to understand our present situation. Society is not a stagnant entity. This is the reason analyses that make judgments based on the present, without concern for the past nor realization that the present is evolving into its future, leave much to be desired. Professor Richard Bawden, in the keynote address at the 1990 meeting of the American Dairy Science Association declared that '... there is a host of crucial issues about reason and forms of logic that need to be investigated if any mindful investigation of new perspectives on agricultural and rural development is to be achieved' (1991).

THE DIALECTICAL APPROACH

Perhaps what we need, to appreciate the contributions of family agriculture, is to increase the scope of our logic. An alternative form of logic was elaborated and has sporadically received attention since the time of Socrates and his predecessor, Heraclitus, in ancient Greece (Kainz, 1988). This method, known as *dialectic*, speaks of relation and movement, of processes and of transformation. Analyses using dialectic, as it has been developed by researchers over the centuries, examine totalities in what today is often referred to as a 'holistic approach.' It is a very useful tool for understanding change, a method of focusing on a moving target. Using this logic it is declared that you cannot swim twice in the same river. This is true for two reasons: first, the river in which you swam yesterday has now flowed on down-river; and second, you are not the same person today that you were yesterday. While this might appear unnecessarily complex for a discussion of family agriculture, it seems so only because we are not accustomed to its systematic use. We are all philosophers – differing only in the conceptual tools that we have in our tool boxes.

Some analysts, principally in the developed countries, have shied away from dialectic because it was used by Karl Marx in his theory. But many others, before and after Marx, have applied dialectic to enhance perception of changing situations. My suggestion is that we keep our minds open to take advantage of any resource that is useful for the understanding of family agriculture and its transformation.

Dialectic has recently been used by physical scientists (*The Dialectical Biologist*, Levins and Lewontin; *Holistic Health and Biomedical Medicine*, Lyng) as well as social scientists (*Towards a New Political Economy of Agriculture*, Friedland, Busch, Buttel and Rudy; *Dialectical Societies*, Maybury-Lewis; *Dialectical Thinking and Adult Development*, Basseches, and many others) to re-examine topics that are undergoing great change.

The first point of importance in understanding dialectic is recognition that change has been, is, and will be occurring. Any viewpoint that does not consider this movement can be valid only for a limited time. Two years ago in a workshop with a group of farm and ranch managers in central Brazil it became obvious to us that the form of logic used in our day to day decisions was not adequate for the pace of change in import prices, government policy modifications and market trends. By applying some dialectic new ideas were soon forthcoming and the situation was put in a more appropriate and transitional focus. A year later at their next meeting several managers were still accusing others of not using enough dialectic.

A recent article on family agriculture in England provides another good example of ignoring the related sides of a question: the author goes into great detail to emphasize that British agricultural production must keep up with the demands for cheap food and more independence from food imports. He then 'admits' that the larger than family farms are better equipped for this task due to their size, higher levels of capital and even the superior educational backgrounds of their operators. These farmers were also seen to have an advantage over the smaller family producers because of economies of scale, more land so that they can make more efficient use of modern machinery and the production system modifications necessary for the application of other innovations. The article gives no attention to the total picture as to whether the research decisions that produce agricultural machinery and equipment consider the needs of various sizes of farmers, how government policy affects different sizes of farm operations, what historical factors have created the present situation of ownership, or even how agriculturally related institutions affect the possible efficiency of different sizes of farms. It is this type of sectoral, partial analysis that arrives at fragmented conclusions and makes recommendations that often address the effects rather than the causes of problems.

A PRIMER IN DIALECTIC

The next few pages will review some of the principles of dialectic in order to apply this logic as we examine some of the complex questions facing family agriculture in the later chapters.

The most effective way to learn new ideas is to relate them to ideas and knowledge that we already have stored in our mental archives. This is especially appropriate for learning about dialectic because it is the form of logic which understands everything in relation to everything else.

But before looking at logic, there are several substitutes that we use when for some reason we don't want to think at all. Sometimes thinking is hard work and we look for an easier way. In this case, conventional wisdom, value judgments and loaded words seem to fill the bill. *Conventional wisdom* expresses ideas that 'everyone knows': that the world is flat, ... that with over a billion people in China agricultural prices are sure to go up, ... that it takes three days for a letter to get to Sioux Falls. *Value judgments* 'help' make decisions: '... anybody from that side of the tracks,' '...made in the USA,' '...those Japanese.' *Loaded words* are used to make others think as we do, perhaps even to convince ourselves: '... the *doctor* said so,' '...a real *yucky* guy' (an adolescent contribution), '...the *best* that money can buy.'

To think more clearly we must avoid these substitutes. They lump unlike entities together, mask individual differences and form opinions for a class of objects from a small number of observations.

Since the early days of civilization humans have developed two forms of logic. One is called *positivism*. The positivist sees the world 'the way it is' and will be. Dialectic, the alternative, is the logic of change. An example often used in my classes with students of Agriculture and Veterinary Medicine examines FIDO and the MOAT. If you remember the 'old days' of cartoons on TV, a Saturday morning feature in many countries, you will recall a common name for dogs of the time, FIDO. If you can remember that name you have the first letters of four principles of positivist thought:

Positivism refers to thought that is: Fixed
                                        Isolated
                                        Divided into categories
                                           of permanent
                                           separation, with
                                        Opposing concepts.

Fixed ideas resist change. What is, is. For instance, eternal love. The positivist idea is that the bride and groom will continue to feel for each other just the way that they feel on their wedding day. Positivist thought *isolates* categories of reality from each other. Students specialize in chemistry or mathematics or physics as though these areas of science exist separately in nature. Government leaders speak of agriculture, industry and services as though they are unrelated sectors in society. A frequently used example of the *division* of positivist thought is the concept of the rich and the poor. It is 'common knowledge' that there will always be rich people and poor people in every society. 'No matter what you do, people are just different.' *Opposing* concepts refer to, for example, life and death. For the positivist these are distinct alternatives. Either you're dead, or alive. There's no third alternative.

Dialectical thought is fundamentally different. It is the opposite of positivism. We can remember some principles of dialectic from the MOAT:

The dialectician thinks in terms of: Movement
                                      Opposition as a union of
                                         concepts
                                      Action that is reciprocal
                                      Transformation of
                                         quantitative change
                                         into qualitative
                                         change.

Often, upon meeting someone that we haven't seen for a while we think, 'He has really changed!' Innately we realize that he has not simply changed suddenly, after a period of remaining unchanged. Change is a form of continual *movement*. This is not to say that change is constant (a positivist term); it can occur either gradually or abruptly. A recent study presented on US National Public Radio found that human babies do remain physically the same size for a period of time and then may spurt as much as a third of an inch in one day. The moving river of the swimmer cited earlier is another example.

Opposition, for dialectical thought, is united. It is like the opposite ends of a magnet, with its opposing magnetic poles. But if we break the magnet in half, each piece will still contain the opposite poles – as will each piece if we break it into many pieces. The opposite poles have meaning only in their relation to each other, just as life is intimately related to death. Prune a tree and a new branch frequently sprouts from the stump of the dead one. Life springs from death, organic matter, a loaf of bread, a chicken dinner, and, of course death results from life. The two are bound together.

The *action* that characterizes the world occurs in *reciprocal* form. This is to say that action generates specific reaction. The reaction is related to the action, which is itself a reaction to a previous action. A paid off mortgage on a farm, for example, is a reaction of a farmer to a situation of debt which was a reaction to a need for capital which was a reaction   and so on.

Transformation first occurs in quantitative form. We can measure the temperature of heating water, 80° ... 85° ... these are quantitative changes that can be registered on a thermometer. When we reach 100° a qualitative change takes place. We have something that didn't exist before – steam. The steam occurs because of the quantitative changes that preceded it. Another example comes from the classroom: when a student sits down and the chair breaks, another student may call out with glee, 'Josephine broke the chair!' But the chair didn't change from its new condition to being broken because of its use by a single student. Over time many students sat on it. Each time the distance between the molecules in the steel legs increased (which can be measured, quantitatively). Then one day Josephine sits down. As a result of the *cumulative effects* of the quantitative changes the chair can't support her weight and it breaks. Now we have a qualitative change, a broken chair. From a dialectical perspective, who broke the chair? Everyone who ever sat on it. At each sitting the steel frame, or the bond between wood and glue, separated slightly, quantitative changes. It was Josephine's misfortune to be the one who sat down just at the point of transformation to a qualitative change.

To develop an understanding of these two forms of thought we must recognize, from the outset, that all of us are both positivist and dialectical in the logic that we use in everyday life. Our goal, to be able to think more clearly, is to use these opposite forms *consciously* for more acute perception. Both types of logic have been elaborated in sophisticated forms by philosophers. But these general principles of each are enough to initiate our usage, and to profoundly affect the way we think. How often have we heard, for example, 'My tooth just broke in half on a fresh slice of bread. . . .' It is unlikely that the bread was the singular 'cause' of the broken tooth. The routine acceptance of positivism can create real problems: 'I always go around that curve at 50 miles an hour . . .'; 'I never have needed a PTO shield on my tractor. . . .' Always and never are favorite expressions of positivism.

We have a tendency to think of things in unchanging form. An article on meteorites in *Newsweek* (23 November, 1992), reports that 'All an astronomer need do is spy the pockmarked surfaces of planets and our own moon to realize that it's a cosmic shooting gallery out there. . . . But it dawned on scientists only recently – in the 1940s – that this pinball chaos hasn't stopped.' It is difficult for us to perceive the possibility of rare occurrences, or of the slow, evolutionary change that is barely noticeable in the space of human life.

Dialectical logic explains that change comes from internal contradiction. An often-used example, perhaps from Socrates or one of the other early dialecticians, speaks of the (fertilized) egg: a unity which, given adequate conditions, transforms itself to produce a new being. (If unfertilized it will also suffer transformation – to a rotten egg. In any case it will not remain as it was.) A chick is not an egg, but it came from an egg. Likewise the chick cannot return to the egg, but will be itself transformed into a chicken. Change is caused by something inside that transforms the whole. You can observe this in yourself. Answer the question: I enjoy farm life because. . . . Now answer another question: I don't enjoy farm life because. . . . If you had answers to both questions you have an internal contradiction, whether you are a farmer or an urban worker. That contradiction may gain importance to the point that it provokes you to make changes in your life, or it may be reduced in importance until it disappears.

As parents we tend to be positivist with our children. Any

problems may be, at first glance, because of 'too much TV,' 'not working hard enough,' or 'not taking care of things.' We have a tendency to see the simplest, clearest explanations as being the most true. Actually the evolving truth is generally more complex. The May apple that comes up in the spring in the US is an example. When it is green, it is deadly poison; when ripe it is edible, and the roots have been used for medicinal purposes, recently in cancer research. Is it a 'good' plant asks our positivist curiosity ... it depends. That's the answer given by dialectic to most questions.

While positivism sees history in terms of cycles, dialectic is more profound. History is relevant, but it's not really possible to return to a previous situation. Dialecticians speak of a *spiral*. The 'return' occurs to a similar point, but at a different level, in terms of other aspects. When the combine is gotten out and greased up for use, the best indicator of how well it will run is how it ran last year. On the other hand, dialectically we need to be prepared for other processes and transformations that are occurring. Michael Basseches (1984:21) explains that

> ... *from the dialectical perspective, what might otherwise be viewed as fundamental elements of existence are instead viewed as temporary forms which existence takes, and what might otherwise be viewed as interactions of fundamental elements are instead viewed as fundamental processes of change through which these forms of existence emerge.*

A fairly painless method for understanding new information is to relate it to our present knowledge and beliefs as suggested above. But this shouldn't exclude the acceptance of unfamiliar information that can't be linked in a 'blanket' relationship. In 1992 the Municipal Zoo in Columbus, Ohio, received two giant pandas on loan from China. The considerable media coverage of the event included an article in the capital city newspaper which asked the question: 'Is a panda a raccoon or a bear?' The forcing of new information into categories of our stored knowledge limits our possibilities for innovation.

A similar example comes from the *New York Times* about the same time. The article announces the discovery of a bird, the hooded pitohue, of New Guinea 'whose feathers and skin are laced with a potent toxin.' Although insects and reptiles are commonly known to have 'chemical defenses' against predators, it came as a surprise to scientists to discover a bird so equipped: '... birds were thought to rely on quick flight to escape.' Simple,

fixed answers, until confronted with divergent information, tend
to be accepted as truth.

We can find recognition of dialectical processes, however, in
the information that circulates all around us. Sometimes it is
rejected. Roger Penrose, in his recent bestseller (1989:115),
criticizes the dialectical theorists in mathematics, saying that:
'... to have a concept of mathematical truth which changes with
time is, to say the least, most awkward and unsatisfactory for a
mathematics which one hopes to be able to employ reliably for a
description of the physical world.' But often it is not that fixed
knowledge is incorrect, just that it is simplistic. Holding on with
such determination to present truths keeps us from attaining a
more complex grasp of reality. Other times the dialectic is
implicitly accepted and used to justify the situation: 'The man the
soldiers are looking for no longer exists. Now there is only a
Sioux named Dances with Wolves.'

The sharpening of our perception is suggested as a goal for all
of us in order to detect the relations that the food and
agricultural sector has with other sectors in society. If we are able
to 'flexibilize' our reasoning processes we can make new what, by
traditional reasoning, appears less than hopeful. More flexible
logic is also useful for our personal lives as every day is a new
day. Family relationships, work responsibilities, and even
recreational activities can become more purposeful and exhilar-
ating through the perception of dialectical process. In *The
Dialectics of Social Life*, Robert Murphy (1971:89) explained
that everyone need not be a trained philosopher: '... but only
that one is willing to draw upon their [the philosophers'] better
insights and allow oneself to transcend the limitations of the
given and the obvious. Our task should be understood as an
attempt to break out of a closed system of thought and not to
adopt one.'

How can we escape ingrained routines of logic to think more
profoundly? One suggestion is to remember the principles: FIDO
and the MOAT. Think how they apply to the subject at hand.
Ask yourself, for example, 'How is this a process and how is it
relating to other processes? How are my ideas fixed so that they
exclude new possibilities?' We may discover new frontiers for
family agriculture that will link with other sectors in society,
transform traditional industry to increase the use of agricultural
products in innovative ways and improve the lives and opportunities of farm families.

To say that farm prices are low, for example, has little meaning in itself. They are low, for those who apply dialectic, *in relation to* something else. This may be past price levels, it may be in relation to costs of production, to urban incomes or other indicators. How does this more flexible way of thinking apply to the social role of family agriculture in society? The satisfaction that a family receives from its agricultural pursuits, the possibility for family interaction, the relevance given to independence of decision making, closeness to nature, and other aspects of rural family life are associated with how others in society value these attributes. Is the farm family admired? Is there symbolic recognition of the contribution of family agriculture to society? When financial crises occur in societies and difficult decisions must be made, how important is the gap between what a family could earn in urban salaried positions versus farm income? These comparisons of life in different sectors of society are indicative of the relationships that often subconsciously influence our thinking.

George Beckford, Professor of Economics at the University of the West Indies in Jamaica, has studied questions of poverty and economic development intensively. He declares that even underdevelopment is a process. If poor countries were left to their own means, without vested interests attempting to prevent change, development would naturally occur. Dialectical analysis offers new insights into how and why processes occur as they do. Relations have great explanatory power. The 'boss cow' may well give more milk than others in the herd because she's able to control more space and eat more at the trough (unless she's unfortunate enough to be in a modern computerized feeding system where this is controlled). Rudolfo Stavenhagen, a Mexican social scientist analysing the Latin American situation declared essentially that the poor are poor because the rich are rich.

Many examples of relations can be found among plants: problems of insects in home gardens and orchards in Brazil are confronted by planting tobacco close to the vulnerable plants. At times of insect invasions you can count the dead bugs killed by the nicotine of the tobacco leaves, under the plants. For more intensive application a 'plug' of tobacco is soaked in water and this tobacco juice is sprayed on infested plants. Farmers in India and other parts of Asia spray a mixture concocted from the bitter pods of the Neem tree on plants to discourage insect pests (Pretty and Howes, 1993). The ecology is thus modified in a non-toxic, low cost manner, and it works. Such examples are typical of crop and livestock methods used traditionally by families who understand the relations among their work objects. On a global basis it can be noted that the objective of agriculture has most often been a sustainable production with low inputs and costs, than it has been profit. The implication is that overt profit-seeking behavior, principally observed in the developed countries, occurs based on different socioeconomic goals that have evolved from the relations among the various sectors of society. This is a situation of critical interest, to which we must return later in our discussion.

Establishment thinking

We speak of dialectic in relation to positivism, the form of logic that is called logic, as if there were no other in western society. It is explained by Lawrence Busch as being used to '... "unify" scientific enquiry to provide a coherent, integrated understanding of nature, including human nature and the nature of societies.' (Joergensen, cited by Busch *et al*, 1991:35). While this may be a worthy goal, it results in fixed 'models' and 'methods,' isolated from the effects of surrounding relationships. The result may be the formulation of one perspective as opposed to some alternative which is actually related to it. Although positivism appears complex, it is the predominant form of logic in the western countries. We learn it from parents as well as our formal education. It is also very comforting. There is no question when your mother says 'If you go out without your jacket you'll catch cold,' or when the newspaper says 'Vitamin C protects you from colds,' or even when the doctor says 'Colds are caused by germs.' It feels good to possess such knowledge. But can all of these 'causes' be valid? The difficulty comes in not being able to relate various causes – and realize that there are usually multiple and varied causes – for the same effect. Feyeraband, cited by Busch *et al* (1991) states:

> *A more problematic aspect of the positivist view of science is its ideal of science as value-free, or at least free from non-scientific values. Science is a human enterprise and all human enterprises are marked by interests, desires, goals, motives, and ideals, and to claim that science is, in essence (ie, ideally), practiced with no regard whatsoever for these desires is to posit a scientific ideal that is practically, if not logically, impossible.*

A dialectical response

From a dialectical point of view we have seen that change is caused by internal contradictions. When a dairy farmer declares that his cattle don't like 'greenchop' during the dry season because it causes itching, he is telling more about himself, who has to cut and chop it, than about his cows. A university department of Agronomy with no laboratories, that explains that students aren't well prepared because they aren't interested, isn't very convincing.

The perceived trade-off between income and family leisure; the prospect of being able to give a son a calf, but not a car; the job security; distance from health facilities; freedom from a time clock; all are items internally valued by the family farmer as having more or less significance. Just as in the chick and egg example mentioned earlier, any of these items could create an internal contradiction and generate dissatisfaction with present situations. The ensuing process might modify our relations with others, and in turn be modified by the effects that others have on us.

Dialectic also takes reciprocal action into account. While agricultural policy, market prices and interest rates are all external factors that have hurt many farmers, internally, how have the farmers responded?

The objective of this discussion is to establish a basis, using innovative logic, for the recognition of the value of family agriculture and to attempt to understand its present evolution and future in relation to aspects of the socioeconomic situation which have not previously been confronted. Of course this recognition is relevant not only for the family farmer, who already has an awareness, but most of all for the rest of society, which has not had the benefit of experience (beyond some memories, perhaps) in the formation of opinion regarding some of the continuing needs of society and the relation of family agriculture to them. For the majority of the population agriculture is remembered basically in terms of food production.

Modern agricultural goals in developed countries, which are also proposed for global use, tend to speak of ecology, sustainability, and of security from human need, among other objectives. It is not difficult to perceive that family agriculture will have additional role responsibilities as it represents a structure of great promise for the realization of such goals.

In the following chapters we will try to apply some dialectical analysis to these goals. If necessary, refer to your Primer in Dialectic in this chapter.

4

Ecology and Agriculture

The terms used in a discussion of agriculture become more complex as we begin to realize the many relations that are involved in a field of production that is absolutely necessary to everyone. Ecology, the relations between an organism and its environment, used to be a subject that we learned about in biology class. From this evolved human ecology the relation of the individual with the institutions and organizations of his or her social environment. US Vice President Al Gore, in his recent consciousness raising book (1992:34) points out the dialectical aspects of the subject:

> ... the only way to understand our new role as a co-architect of nature is to see ourselves as part of a complex system that does not operate according to the same simple rules of cause and effect we are used to. The problem is not our effect on the environment so much as our relationship with the environment. As a result, any solution to the problem will require a careful assessment of that relationship as well as the complex interrelationship among factors within civilization and between them and the major natural components of the earth's ecological system.

As studies became increasingly specialized Miguel Altieri (1990:16) suggested that the farm family and agricultural development be studied using *agroecology*. Such an approach would be '...more sensitive to the enormous variations in ecology, population pressures, economic relations, and social organization prevailing in the region.' This orientation examines, for example, farmers' reasons for making changes in methods and production levels in terms of the various factors of their situations, and demonstrates the importance of family agriculture to moderate the strain on land and other resources which reflect, with time, on the society at large. For successful analysis using agroecology much information is needed from farmers. The

effects of governmental controls, price changes of farm products
that result from fluctuations in currency values between countries, enormous personal loans and financial responsibilities,
distance from community services, risk due to climatic factors,
market instability and many other aspects of production and
farm life are problems not faced by families in other sectors of
society. In the developed countries the leadership of the
agricultural sector has had some success in bringing this situation
to public knowledge. But the minute proportion of the population in the agricultural sector means that few people know
farmers personally. In the developing countries, where farmers
have frequently not had the benefit of formal education and
organizations for their participation are nearly nonexistent, an
approach such as agroecology is more difficult. Even scientists in
these countries who study agricultural topics are generally not
from rural areas, may not be acquainted with agricultural
producers and base their research on secondary information of
what the agricultural sector needs. In Peru an extension worker
told me that scientists working with potato storage problems
discovered, only after devoting considerable research to a solution for losses, that producers were not as concerned with rotting
as they were with problems of the potatoes sprouting and losing
weight in storage. Lack of responsiveness to the realized needs of
farmers results in innovations not being readily accepted,
research resources wasted and the felt problems of farmers left
unanswered.

As agriculture decreases in economic importance relative to
the industrial and service sectors, less public funding is available
for research. The studies of private firms in developing innovations – seed, fertilizer or machinery – are, of course, conducted
with the objective of increasing the share that their products hold
of the market. From this orientation, sensitivity to the ecology
and details of production that are not directly related to use of
their products may receive little or no attention.

Where there are common resources; national parks, rivers,
common pastures, communal farms, we are even less capable of
effective management. The Club of Rome reported twenty years
ago that nations have never developed effective structures for the
adequate protection of these resources. Since then the situation
has worsened. Common lands as well as private properties of
farmers under serious financial stress have been excessively
exploited for survival and the realization of short-term objectives. Every year new studies find that major cropland areas are

eroding faster than natural processes can replace soil nutrients. A report issued by the World Resources Institute (1992:1) states new estimates of the world's leading soil scientists: more than 1.2 billion hectares (2.68 billion acres) of 'vegetated land' – an area as large as China and India put together – have been significantly degraded since World War 2.

Figures 1 and 2 illustrate differences between the natural ecosystem and a 'human subsidized agroecosystem.' In the natural system the principal source of energy is the sun. The addition of human subsidies in the form of chemical and mechanical innovations is shown in Figure 2. It has been traditional for farmers all over the world to seek innovative methods: new plants, improved methods and techniques and modern machinery for adaptation and integration into ongoing farming systems. But the situation portrayed in Figure 2 has been carried to an extreme with integral dependence, in some regions, on these outside subsidies. There is thus implicitly less importance given to the knowledge and skill of the farmer than has been the case. He or she has more resources to outsmart the problems. Actually the 'traditional' knowledge of agricultural production in all parts of the world, at the same time that it is being passed down

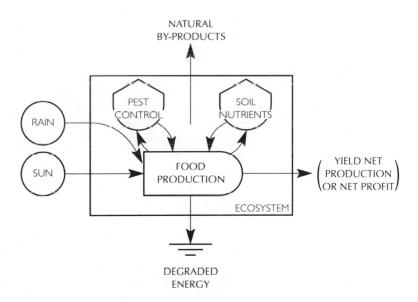

Figure 1 *A natural unsubsidized ecosystem*

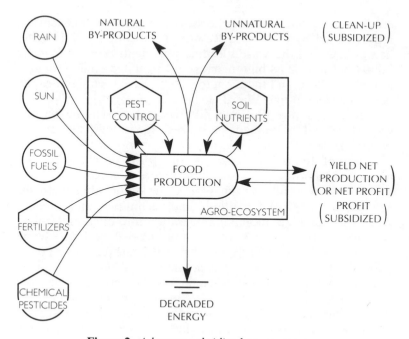

Figure 2 *A human-subsidized agroecosystem*

Figures taken from G W Barrett, N Rodenhouse and P J Bohlen, 1990

through the generations, is continually being adapted to changing conditions, incremented, and reformed. The possibility of more intensive management applied to the 'human subsidized' situation does still offer hope for the recognition of traditional knowledge and skill. The targeted use of small doses of both pesticides and fertilizers in some places represents a modern example. Steps to stimulate the recycling of nutrients within the system, minimizing wasteful inputs and optimizing, rather than maximizing, nutrient outputs are renewed methods of improvement.

To consider these and numerous other possible modifications in the context of the larger ecosystems, we would examine the complex of nutrient cycles, the dynamics of population change in numbers and demand, and variations in the quantity and types of energy that would be required. The result could be better human health as well as economic well-being. This is not to imply that standardized programs should be developed for national implementation. The need is for examination of local, generally family, systems and then their relations to the larger system. This orientation has been advocated by Chambers, Pacey and Thrupp in their book, *Farmer First, Farmer Innovation and Agricultural Research* (1989). It has also been recommended by Conway and Barbier (1990:32), who have noticed that 'Mismatching of technologies is particularly apparent where technological packages are applied on a large scale, in the belief that the natural resources they are intended to manage are uniform.' While guidelines must be established at national levels, investigation of ecological well-being is similar to the principles of democracy: it starts at the grassroots level.

We can begin with field practices. In the Latin American countries 'succotash' planting, whether to cultivate beans under corn, has frequently been discussed. In Mexico, farm level research found that yields could be increased by as much as 50 per cent by interplanting 'maize' with beans and squash (Gliessman, 1990). While the yield of any one of the crops was less than it would have been if planted alone, the total harvest was more and was more stable from year to year. Corn and beans or rice and beans are the staple food crops of much of Latin America. The traditional varieties, while well adapted to local conditions and thus low risk, are also low in yield.

Varieties and techniques developed by the international 'Green Revolution' program to genetically improve indigenous crops and increase yields have met this objective, but also are more demanding of fertilizer, pesticides and water and don't always have the characteristics considered desirable for local dishes. Professor S Satish, of the Administrative Staff College of India, reports that many rice producers there prefer to sell production of the new varieties and eat their traditional grain. There are stories that the high yielding corn in Mexico is inferior for making tortillas. Some of the techniques used with the improved varieties have also raised questions. Examining agriculture in Asian countries that adopted Green Revolution seeds and methods, Edwards and his colleagues (1993:XVII) explain that 'Growing crops in continual monoculture or biculture leads inevitably to much greater pest, weed, disease, fertility and soil erosion problems than the mixed cropping systems common in developing countries.'

But the high yielding grains increase local activity. Roads are improved, marketing systems incremented and private firms intensify their sales programs to help farmers spend their new incomes. Families make purchases, frequently entering into credit relations with firms and banks and begin to make production level decisions based on their needs for cash incomes. Crops that had been consumed are sold as farmers seek to make payments. These occurrences are indicative of the changing relations among humans, animals and plants, and even among plant varieties, and result in altered social relations as well.

THE APPLICATION OF DIALECTIC

In the last chapter the possibility of using dialectical thought was suggested to increase our holistic objectivity in understanding agricultural problems. One of the principles of dialectic is *transformation*. Transformation, as a form of change, may begin quantitatively and evolve to produce qualitative effects. Remember the example of the little girl who sat on the chair just as it was at the point of qualitative change (Chapter 3)? Her (positivist oriented) colleagues reported to the teacher, 'Josephine broke the chair.' The cumulative effects of the quantitative changes couldn't support her weight and the chair broke. This is to say, there was a change in the qualitative state of the chair.

What does all this have to do with farmers borrowing money? Using dialectic we can see how at first money is borrowed, and that quantity changes with time. If all goes well the money will be repaid and the qualitative change, or transformation, will be quite positive: a farm paid off, the family confident in its success, a good credit rating, a satisfied bank, etc. If, on the other hand, there are problems related to the process as the quantitative changes (modifications in value) occur, the result may be a qualitative transformation that is not so positive, worse than a broken chair. The 'debtor' will suffer various types of transformation which, it is important to understand, don't happen simply because of that last loan. They represent the accumulation of all of the quantitative steps of the process. Implicitly we recognize this when we refer to 'the straw that broke the camel's back.' But by knowingly applying dialectical thought while we are 'in process,' we may be able to avoid that last straw. Certainly the individual who experiences serious problems with debt will be qualitatively transformed and will approach future financial problems and opportunities from a more knowledgeable point of view (commonly described as 'once bitten, twice shy'). Processes occur in quantitative-qualitative sequences. Recognition of this movement sharpens our ability to control our personal processes as well as to work in harmony with the ecological transformations of nature.

Families seek more than profit

Producers manipulate the ecology of their farm situations to stabilize production from year to year to make better use of their resources and to meet personal needs. Manioc or cassava (tapioca) is a starch crop grown in many tropical and subtropical countries. The tubers are easily stored (left in the ground until needed) and the leaves are a rich source of protein, used principally in Africa where other protein sources are scarce. There are many varieties of manioc which have varying characteristics. Some have high levels of prussic acid which the indigenous people of Amazonas press out before consuming the manioc. The acid is used to marinate the somewhat rubbery duck meat. Some types of manioc cook more quickly, good for when food preparation time is limited; others are more resistant to drought, soil acidity or local pests. It would be risky to plant only one variety in the face of possible climatic or pest problems. By raising several varieties families maintain stable production and achieve multiple goals. Squash is another example of a crop that families plant different varieties of, for varying goals. The producer, through knowledge of his ecological situation, is thus able (within limits) to maintain the desired levels of production.

Writing on the importance of maintaining crop genetic diversity, Laura Merrick (1990:4) observes that, especially among small producers, differing varieties of a crop are selected for planting, either all together or in different fields. They

> ... are commonly distinguished by farmers by such characteristics as color of different plant parts, yield, culinary quality, resistance to diseases or pests, storage properties, altitude for optimum growth, and growth habit. Unlike selection and breeding criteria applied to modern crop varieties, yield may or may not be an overriding concern in the selection of traditional varieties by farmers.

Sometimes individuals take steps to modify production methods and improve the ecology of the situation even though it goes against the social norms. In Europe, agricultural producers are rethinking their tradition of maintaining clean crops. Sugar beets sprouting in a clean field are often devoured by insects which, it's been found, would be just as satisfied to consume the local weeds. Weeds also provide soil cover until the beets attain their growth.

Social norms have dictated that 'good farmers' keep their fields clean. But the weedier fields end up with better stands of beets and less erosion.

The nations of the European Union have provided funding for 'environmentally sensitive areas' since 1985. Through management agreement programs farmers are paid to retain traditional agricultural practices in order to maintain biologically important habitats and traditional landscapes. Contracts are usually written for five year periods. Other programs of EU countries include the following:

- The United Kingdom Countryside Commission provides premiums to farmers who implement environmentally preferred management practices on set-aside lands.
- In almost all German townships, farmers can receive payments to protect native plants and animals by leaving some meadows uncropped and avoiding the application of certain fertilizers and pesticides on grassland and on the edges of fields. Most such programs are managed by state level government.
- Sweden and some other European countries provide investment aid to farmers who upgrade manure storage facilities.
- Portugal, as part of its program to reduce soil erosion, offers subsidies and loans to farmers for reforestation and the establishment of permanent pastures.
  (*Taken from the World Resources Institute, 1992:106*).

Specialization versus diversification

The crop-livestock diversification of agricultural production, which traditionally offered alternatives to maintain an ecologically satisfactory situation, has been nearly lost in some areas due to the specialization of production into intensive crop or livestock operations. This type of separation of production leads to ecological problems. MacCannell (1988) has noted that the new industrial farms of the southwestern United States are 'qualitatively different' from family operations. Not only are they larger but they also have higher levels of capital investment, off-farm labor requirements and are more integrated with other economic interests in society. They depend heavily on innovations that respond to their particular characteristics and have the political influence to persuade researchers to investigate their problems. Unfortunately the resulting innovations are less useful to mainstay family agriculturalists and may be quite demanding on society. MacCannell (1988:28) reports that 'Today, industrial agriculture has a near monopoly on fresh water supplies in the Sunbelt.' In the states of Florida, Arizona, Texas and California, between 75 and 90 per cent of the total water supply is used for large scale crop irrigation. The small part that is left must do for all other farm, industrial and urban-domestic requirements. These industrial farming operations, if they are to maximize profit, are unable to show sensitivity for long-term issues and ecological questions. They are a little like some US farmers at the turn of the last century who boasted that they had 'worn out' several farms, and soon moved on to urban employment.

The effect of farm size on ecological methods varies

In most countries trees and shrubs are useful for ecological objectives. They reduce wind erosion and bring water and soil nutrients up from depths where they are unavailable to agricultural crops. The leguminous species contribute to the fixation of soil nitrogen, and organic matter is formed by the decay of leaves and rotting branches and roots. This process also improves soil porosity, reducing runoff water and the corresponding erosion. The result is improved nutrient recycling in the system.

Where trees are incorporated into an agricultural system there is less need for forage production, thus reducing tillage. Leucena is a subtropical example that, being leguminous, is quite nutritious as feed for livestock. Mulberry is cultivated for silkworm production. Of course firewood is a widely used source of energy and fruits and nuts also increase the desirability of trees. Problems arise for small farmers in tree production, however, because it involves land displacement ('I could raise a bushel of wheat where that tree stands') and the difficulty of establishing seedlings on land that is periodically grazed. This is especially true for leguminous species, preferred by cattle – and goats will prefer almost any tree to pasture. Thus, while small farmers have appeared uncooperative in some tree planting projects, the difficulty has often been that they just don't have adequate conditions to protect young trees.

In some surprising research results from Ethiopia (Kirkby, 1990), it was found that yields of cereal grains planted under the leguminous Acacia trees actually were 55 per cent above the yields in adjacent open field plantings. The Acacia tree is a deciduous, medium-sized tree which is often planted for firewood in areas of limited energy resources because of its rapid rate of growth.

THE INFLUENCE OF SOCIAL NORMS AND ECONOMIC INTERESTS

Perhaps the topic of initial importance in any program of agroecological planning is the identification of social norms that regulate human activities, how they develop, are maintained and are transformed. In Africa the social group of most influence is the family. Individual needs and interests are subordinated to family needs and little regulation (beyond paying taxes) has been imposed, until recently, from higher levels of government. Since land in areas of slash and burn agriculture is considered to be the common property of village residents, there is generally no family protection of specific plots for future generations. But with increases in population there has been government pressure on local farmers to produce more food. The fallow period between crops has been reduced from twenty years or more to sometimes only three years. In this short time the soil is unable to accumulate the necessary fertility to produce another crop. Food production, on a per capita basis, has thus fallen in recent years in most of Africa (Conway and Barbier, 1990:20).

To understand from a socioeconomic point of view the ecology of African agriculture, it is necessary to consider the interests of ex-colonial powers and other developed countries in importing raw materials: rubber, coffee, cocoa, peanuts and others, for processing. As prices of these crops rose over the years arrangements were made for land consolidation and highly subsidized inputs were offered by the developed countries for more intensive, large scale production of export crops on the most productive land. Local capital resources increased and the traditional food crops were abandoned for 'French bread' locally baked from imported wheat, and similar items considered exotic and thus superior to native African staples. Then when the prices of export crops dropped during the 1980s, cash incomes were reduced and people sought their native food again. But it was difficult to return to the production of food. Rubber, coffee and cocoa are tree crops that require several years to establish. This generates reluctance to cut them, to return to the production of annual food crops for local markets. Peanuts and other crops planted for export involve cash inputs that price them out of the budgets of local consumers. There has subsequently been an increase in welfare food shipments, and in hunger. To regain adequate levels of local subsistence in the African countries would require major changes in land use patterns, production systems, population settlements, capital investments and international trade. As these changes are avoided, delayed or inadequate, suffering from the ecological imbalance will increase.

At the other extreme, the high social, environmental and resource costs of large agroindustrial units in the developed countries, as cited in the US Sunbelt cases above, are beginning to be realized by the general population. The expenses generated by these mechanized operations requiring vast quantities of resources, hired labor and specialized research are based principally on the economic objective of cheap food and involve changes in the social norms of the areas involved. Conway and Barbier (1990:30) sum up the concern:

> Developing countries, however, are recognizing that they cannot afford the technological investment.... At the same time in the developed countries it is becoming increasingly clear that many of the technological solutions, for example use of pesticides and artificial fertilizers or 'industrial' livestock production, have high and unexpected costs and, more importantly, are themselves in many respects not sustainable.

From the point of view of human ecology, farm families have also been affected by rural-urban change. As cities have developed, the ideology of urban superiority, with its unbridled consumption, has threatened the rural family traditions. In some countries there has been a recognition of urban biases in rural elementary school materials and programs. Efforts are being made to develop specialized materials that reinforce a positive orientation and the advantages of rural life. On the other hand, a study in southern Brazil found that parents felt materials tailored for rural students were inferior and undesirable. They preferred that their children receive the basic skills required for leaving the rural area and integrating into urban society.

This shift in the human ecology of rural areas began as the urban populations grew in the developed countries, which had industrial infrastructures for their absorption. As the trend spread to developing nations cities were referred to as 'swollen,' showing 'unhealthy' size. Basic services became unsupportable burdens to municipal budgets. The rural-urban migration in most countries took place in a stepwise manner. Families first moved to a nearby town, perhaps where they had relatives who could instruct them in urban ways: how to find a job, where to look for advantages, how to avoid legal problems. Many of these families subsequently moved on with their newfound knowledge, to larger, industrialized areas. Developed countries observing this process considered it natural – it had also occurred in their industrialized areas. They offered aid to alleviate the problems that at the same time sanctioned the process. The result has been a profound ecological imbalance.

Only as governments recognize the need to create millions of jobs in the next few years is it perceived that policies that reduce the size of the farm population have resulted in a trade of independent underemployment on the farm for dependent unemployment in the city.

One exception to these trends, as reported in Chapter 2, has been Hungary. With the opening up of the ex-Soviet bloc, visitors are amazed to see the Hungarian industrial development without high levels of urbanization. Although there is underemployment in rural areas, workers prefer to drive to the urban factories from their farm homes than relocate to the city (Szelényi, 1988). Still the entrenched ideas of the developed nations persist. Recently a returned visitor from Hungary commented in a US presentation on the problems that he had observed: 'Hungary still has 20 per cent of its population on farms while the United States has only 4 per cent.' But from an ecological point of view the Hungarians, with these procedures, may be more able to maintain the ecological balance that they desire.

Research results from Chile (International Food Policy Research Institute, 1990) report that a 10 per cent increase in agricultural prices that occurred in 1963 resulted in a 4.4 per cent increase in output four years later, which rose to a 10 per cent increase in 10 years. This increase was obtained by raising capital investments approximately 10 per cent (slightly less than output increases) together with 15 per cent more labor. Labor use continued to rise to 23 per cent in 20 years while capital inputs stayed nearly the same and output increased slightly. What happened? The rural population responded to what was considered, implicitly or explicitly, an ecological problem. With increased economic opportunity, food was produced using more labor, more sons and daughters staying at home, and less 'high capital inputs' were required.

There was a somewhat similar occurrence in the United States during nearly the same period. Tweeten (1989) reports that a

total of 327,000 persons under 35 years of age operated US farms
in 1969. This number increased by 8 per cent to 356,146 in 1982.
He explains the increase as being due to 'credit on concessionary
terms, leasing arrangements, and, more importantly, assistance
from parents.' Tweeten then makes a statement that reveals a
good part of the secret of how and why family agriculture is
capable of handling the responsibility for feeding the nations. He
says: 'Without technological, managerial, and financial assistance handed from parents to their children who farm, the family
farm would nearly end in a generation.' Then, with optimism
(1989:18):

> That assistance will continue, however, assuring a supply of
> new family farm operators in the foreseeable future. The ratio
> of retiring and deceased operators to farm youth is increasing.
> Hence, would-be young operators will have a higher chance of
> getting into farming during the next two decades than during
> the 1950s and 1960s unless the pace of farm consolidation
> accelerates.

The probability of the rejuvenation of family agriculture in the
US should be positive for 'agroecology.' Assistance and support
from parents also integrates agricultural production into an
inter-generational operation more similar to that found in the
rest of the world.

SUMMARY

We have seen that the family is an ecologically responsible
structure that promotes a balance between the continuity of
agricultural production and the protection of renewable and
nonrenewable resources.

Although research has made it possible to increase production
through the industrialization of agricultural processes, it has
sometimes resulted in ecological imbalances and the destruction
of resources. This situation has not been fully realized as it
continues to produce profit, and ecological concerns are ridic-
uled or put off to be considered at a future time. Some of the
successes of agroecology that we have discussed demonstrate the
need for intensive labor and personal concern in agricultural
production. The planting of several different varieties of a crop,
for example, to assure stability, is not a production method of
industrial agriculture. Planting grain under trees, in spite of
increased production, imposes limitations on the use of large

agricultural machinery. The restricted application of inputs to reduce pollution and cut costs requires detailed decisions and long-term knowledge of land characteristics. These few examples point strongly to a continued need for the family as the most effective structure for agricultural production.

To examine ecological questions and family agriculture over time we turn, in the next chapter, to the sustainability of agriculture.

5

Sustainable Agriculture

Sustainability, as a goal for agricultural production, means different things to different people. There is some confusion as to whether ecology and sustainability can be used interchangeably. I would suggest that the relation between the two concepts is similar to that between hygiene and health. Hygiene is an immediate and principal component. Health, like sustainability, is more of a general, long-term objective. Some speak of increasing the efficiency of agriculture to sustain a growing world population. Others speak of minimizing the use of nonrenewable resources without recognition of world population growth. Obviously neither of these orientations would be adequate to plan for the production of food and fiber for future generations. What stands out when we attempt to discover the forms of agricultural production that could be sustained for long-term production at adequate levels to meet our needs is that many related factors are involved.

Efficiency is, without doubt, important. But the term is more associated with profit than with the possibility of long-term production. Efficient use of capital, land and labor generally refers to high levels of production in relation of each of these factors as they are applied. To produce the food and fiber that are essential to the well-being of the citizens of each country, there is need for efficiency in the use of capital (it is limited and could be used for other social needs) as well as in the use of land (it is also limited, more in some countries than others, and defines levels of producer income). Labor, on the other hand, is not so easily qualified. If we accept the goal of minimizing labor per unit of output, mechanization is necessary, and probably debt and dependence on government and financial institutions. It becomes more difficult for sons and daughters to remain on the farm. In this situation, as it presently exists, the sustainability of family agriculture itself is called into question. The farm population dwindles to a small proportion of the national population.

Political power is substantially reduced and alternative forms of 'more efficient' production emerge.

In the developed countries criteria for evaluation of social and economic projects have increasingly stressed the number of jobs that will be created. But the importance of minimizing labor in all profit oriented undertakings creates an impasse. Our children leave the farm to increase labor efficiency, only to face unemployment in the urban area. Studies in the developing countries, where capital resources are often severely limited, however, show that the most feasible means of increasing the productivity of land or of capital is to increase the labor applied to the process. Small farmers consistently demonstrate higher indices of productivity when measured in this manner. And according to Pretty and Howes (1993:63) we should not worry that sustainable agriculture will necessarily mean reduced income:

> *Until recently, it was widely assumed that sustainable agricultural practices could only bring lower returns to farmers. They were thought to be low-input practices. It is becoming increasingly clear, however, that integrated farms can match or better the gross margins of conventional farming, even though there is usually a yield per hectare reduction of some 5–10 per cent for crops and 10–20 per cent for livestock.*

Those who speak of natural resource conservation are concerned about the loss of soil productivity from erosion, plant nutrient losses and shortages of nonrenewable resources, as well as surface and ground water pollution from pesticides, fertilizer and sediments (Parr *et al*, 1990). As additional elements of sustainability are examined it becomes even more clear that there are complex relations among the many factors and considerable contradiction as attempts are made to modernize agriculture and sustain the farm family.

PROFIT AND PRESERVATION

Various orientations to production have evolved. While there is stress on differing goals these orientations can, in general, be plotted on a continuum that ranges from total preservation to total profit, as presented in Figure 3. This is not to say that either profit or preservation can be completely isolated from its opposite. In the last chapters we have discussed dialectics and the importance of examining totalities rather than isolated parts. Another principle of dialectical logic is the *unity of opposites*. Just

Preservation	*Profit*
Conservation 'Organic farming'	Industrial agriculture
'Alternative agriculture' 'No till' 'Plow till'	
	High capital inputs
High 'produced inputs' 'Sustainable agriculture'	
'Biodynamic production'	Specialization Mechanization 'Modernization'

Figure 3 *Production orientations*

as it is impossible to break a magnet in two to separate the positive from the negative pole – each piece will itself contain positive and negative poles – profit and preservation are linked in a unity on which points can be located between the two that represent intermediate positions. The use of capital to increase preservation, for example, will reduce profit. Application of capital in the form of mechanization or chemical fertilizers to increase profit will reduce possibilities for preservation. Of principal interest to this discussion is the relation between the two poles. Dialecticians give the example of the slave master and the slave. It is only through knowledge of the relation between them that their differing roles can be understood. If this relation is ignored or unknown, they are just two people. So it is with the orientations of Figure 3. We will perceive the preservationists' orientation by examining it in light of profit, and vice versa.

Unfortunately applications of land and labor don't fit so neatly into this dichotomy. They are related to the process, but in multifaceted ways. Land may be used extensively or intensively either in pursuit of preservation or profit, or anywhere in between. Small farmers often use land more intensively, reaping high levels of land productivity. But this doesn't necessarily indicate their preservation-profit orientation. Labor too can be used for either preservation or profit. The slash and burn agriculture of tropical regions is high in labor use, with the principal goal of subsistence! The orientations listed in Figure 3 include ideas that have become well known in some countries and have generated interest as possible innovations for the future. By the fact that they can all be positioned in the diagram, it is clear that the discussion of sustainability lies mainly on the preservation-profit continuum. There has been some confusion of terms. 'No till' or 'Plow till,' for example, is thought by some to be a synonym for sustainable agriculture. Although it does offer advantages in terms of conservation, it basically represents an attempt to cut costs and may require use of pest control chemicals that counter the principles of sustainability. 'Organic farming' or 'Biodynamic production,' on the other hand, may stress preservation more heavily than proponents of sustainability, producing less than would be possible given the resource base.

There are other aspects of production orientations that don't fit into this representation. A family may use differing orientations on rented than on owned land or may simply hold lands for speculative reasons without giving importance to profit (from production) or to preservation. Indeed the purposive selection of any of these orientations requires considerable knowledge and skill. The figure suggests, however, that capital, when it has been applied intensively, has generally been intended to increase profit.

THE RELATED FACTORS OF SUSTAINABILITY

Plucknett (1990) has explained sustainability in terms of '... the complex interaction of biological, physical and socioeconomic factors.' Among the biological factors he includes genetic resources to be strengthened and maintained, productivity by area and by unit of time, long-term pesticide control and balanced production systems involving both crops and livestock. Physical aspects include soil and water management, use of agricultural chemicals, atmospheric changes and energy consumption. Socioeconomic factors also act to promote or inhibit sustainability, depending on the abilities of governments in the formation of appropriate and timely policy decisions as well as the delivery of credit, inputs and transport. It is, of course, essential that these factors come together to create a situation that is economically feasible for sustainability to become an alternative for farm families. Bawden (1991:236) agrees:

*As agriculturists we need to rethink our fundamental
perspectives on what we actually mean by improvements in
agricultural and rural development. The language of reduc-
tivism and positivism do not entertain the very complex and
dynamic phenomena associated with the quest for sustainable
practices.*

To assess feasibility we need a method of analysis that considers
all of the various factors within an evolving frame of time during
which the importance of each may assume variable levels. Using
the logic of dialectic we may perceive that, over time, the benefits
of sustainability add up to more than the results of other, short-
term orientations.

Dialectic is more than a method. It is a philosophy. The
manager of a dairy farm in Brazil recently provided an interesting example. The farm owner where he worked brought a
friend, a German World War 2 prisoner, to visit the property.
The friend expressed great joy in returning to the rural situation
and told of his own dairy experience in Germany, many years
before. The owner then invited him to spend a week or so on the
farm to become reacquainted with life on a dairy farm.

The visitor agreed with great pleasure and that very evening, after the owner had returned to his residence in the city,
he began participating in the milking. A few days later, looking over production data, the manager noticed that the cows
on one side of the herringbone milking parlor had produced
more milk than they had previously been giving and than the
cows on the other side of the parlor were giving. Feed or feeding methods had not been changed and in any case were the
same on both sides of the parlor. Suspecting that it might be
an effect of the old German's presence he asked him to
change places with the man on the other side. Sure enough, a
few days later the cows on that side had increased production.
The manager then observed the German's procedures. He
noticed that he spoke to each cow as she entered the stall, passed his hand over her, washed the udder with warm water and
paid close attention to how long the milker was left on. These
were steps that his colleague, milking on the other side, considered 'not our way.' If the manager had been less observant
the differences might have gone unnoticed, or been attributed
to random circumstance. His unconscious use of dialectics,
however, made it possible to identify the difference as having
occurred due to an internal change in his milking system.

Another look at logic

Dialectic speaks of change arising from internal contradictions. In Chapter 3 we spoke of Socrates in ancient Greece. It may have been he or some other of the early philosophers who held up an egg to his students. The egg represents a unity. This was expressed many years later by a German philosopher, Fichte (according to Murphy, 1971:88), as a 'thesis'. There are forces opposing the unchangeability of the egg that are inside the egg itself, represented, for example, by the sperm cells, an 'antithesis', from the rooster (this example worked better with students from the rural area). Depending on the external conditions surrounding the egg – temperature, humidity variations and time – this antithesis becomes more or less important in terms of changes that take place inside the egg. Under certain conditions the sperm cell antithesis will grow and develop until it destroys the egg thesis, and a 'synthesis', a baby chick, is the result. We can note from this example that the external conditions are important depending on the internal situation. An antithesis is not constituted by just *any* force that acts against the egg (such as dropping it on the ground), but that force that is inside the egg itself.

In the case of the dairy manager, he might have sought to increase production and profit by modifying his feeding procedures, product sales, or participation in groups that lobby for governmental dairy subsidies. But at least one of the antitheses that affected increases in his production was to be found right inside the milking parlor, in the handling of the cattle. This type of analysis can be employed at various levels of systems. We might compare the sustainability of agriculture in two regions and discover differing internal characteristics which influence production. By the same token, orientations that have been sustainable in the past may not in the present produce the same results. Benbrook (1990:69), has noted that

> ... *the problems faced by American agriculture that could undermine sustainability differ greatly by region, both in degree and character. Many, if not most, agricultural systems currently practised in the United States are likely to remain sustainable for many years to come, albeit perhaps at higher cost to the treasury and natural environment than desirable. Adjustments in cropping practices and technology surely will be needed for most farms to remain sustainable and competitive, but such adjustments will be made.*

Thus it is not likely that one general policy to promote sustainability will be adequate for an entire nation in view of the many factors and conditions involved in the different regions. In an article on 'Conservation Tillage and Sustainable Agriculture,' Rattan Lal and colleagues (1990:219) explain that just in terms of the type of tillage to be used there needs to be consideration of '...soil texture, structure, erodibility, susceptibility to crusting and compacting, slope length, slope gradient, slope aspect and shape, effective rooting depth, plant-available water and nutrient reserves and internal drainage.' They continue to elaborate each of these characteristics and its importance to sustainability. And this is in addition to aspects of rotation, diversification and the many other factors that must be kept in mind.

A review of what has been written on sustainability reveals most concern in the developed countries and in the developing nations where population is increasing rapidly. In the United States and some countries of Western Europe, for example, pressure for high grain yields has altered the rotational systems recognized for generations as being critical to maintain soil fertility. Inexpensive chemical fertilizers have made it appear unnecessary to maintain livestock to recycle nutrients and build up organic matter in soils. The livestock ends up concentrated in feedlots far from the fields, and the manure a waste product to be disposed of.

In the developing countries producers have intensified the use of marginal lands and permanent pastures have become seriously overgrazed in efforts to increase production. Objectives that have created these predicaments, both in developed and developing countries, have sometimes been profit and sometimes subsistence. Governmental policies, rather than alleviating the problems, have often encouraged the maximization of production and productivity, land clearing, draining and rural-urban migration, thus neglecting the future implications for land and water as well as human resources.

Applying logic to define objectives

Basically the hope of sustainable agriculture is for ongoing production to maintain an equilibrium with the changing demands of a growing population in view of the problems of nutrient removal and environmental degradation. The standard response to population growth has been to increase production. The methods applied, however, have generated other problems. It is not surprising that the higher levels of production have nowhere been declared a lasting solution for food supply. Separating the components of agricultural systems that have existed for thousands of years, to create specialized grain and livestock operations, requires innovation that is superior to that traditionally used in all of its effects. While the producers involved may choose to overlook some negative aspects as unimportant, these effects may evolve in the manner of the internal contradiction spoken of above, to the point that they can no longer be ignored. There will be a need for reevaluation. Another fork in the road may have been a better choice.

Advocates of sustainability predict an inevitable return of livestock to the grain producing regions. The interaction of crops and livestock makes increases in soil organic matter feasible and forage crops can be reintroduced into rotations to build up soils, reducing the necessity of purchased inputs. This may call to mind the problems of building and maintaining fences and mowing fence rows... the feeling is that we're going back to old production methods. Here again the logic of dialectic can be of service. The dialectician does not understand events to occur in what others describe as cycles (sometimes 'vicious'). He or she explains, as we saw in Chapter 3, that even as events seem to repeat themselves, aspects of the situations have changed. It is thus more appropriate to envision these events in the form of a spiral (which may be moving upwards or otherwise, depending on perception of the events). Future pastures, for example, may involve short periods of intense use of many small fields. Electric fences that can be easily moved may reduce traditional problems and modify the concept of fencing. Crop innovations that free up labor could make it possible for the producer to enter into some particular aspect of livestock production or other activity. The use of sexed embryos in dairy cattle would increase available resources for other animals by eliminating unneeded bull calves. The return of livestock to the crop farm also gives back to the producer the flexibility by which sales can be diversified, both in terms of products (grain, or meat, milk or eggs, for example) and of time, to meet the family's particular needs.

While we tend to think of questions of soil depletion and future production possibilities as situations that have recently become relevant, Shi ming Luo and Chun ru Han (1990) did an extensive review of literature on Chinese agriculture, going back 7000 years. Their evaluation (1990:299 and 309) is that:

· The development of that agriculture has been seriously shaped by nature and society. Those most widely practised and recorded in ancient Chinese agricultural literature were ecologically reasonable. . . . Several characteristics of the rice-based rotation systems make them more sustainable than other forms of field usage. The water layer of the rice field has a protective and diluting effect to reduce erosion of soil and nutrients during the rainy season. . . . Nitrogen fixation is high in rice fields. Eighty-one species of blue green algae in rice soil were identified [in one province] . . . and many species of bacteria in paddy fields also have a nitrogen fixing ability.

It is interesting to note that one of the texts read by the authors in preparation for this article was the *History of Dialectical Thinking in China (Early Qin Dynasty)* by Ke Fan. They recognized that the way the people think is associated with the ways that they make their agricultural decisions.

Diversification

As agronomy students in the 1950s, we were taught that the only reason to haul manure and spread it on the fields is that it's easier than moving the barn. Fertility was a question of modern, chemical inputs, more effective than the bulky, low nutrient manure. With the movement of feeder cattle to the feedlots manure was farther away and costs of collecting, storing and hauling increased. At the same time chemical fertilizers became economically practical and were used in greater amounts to overcome limits of soil fertility and increase productivity. Feedlot beef cattle, or dairy cattle kept in drylot, however, are chiefly *ends* in the production process. When they are incorporated into diversified systems, they become *means* for the utilization of crop by-products and the production of manure as well. In the absence of diversification there is evidence that the increases from intensive use of chemical fertilizers have limits. Figure 4, as an example, presents three types of soils. Lal and his colleagues (1990:205) explain that

For some fertile soils (soil A in Figure 4), high yields are obtained even with no input. In contrast, soil C does not respond to any level of input. There are vast areas of marginal soils of category C that should not be cultivated. A large proportion of arable land falls under category B and responds to input. The level of input should be judiciously managed for soils of category B.

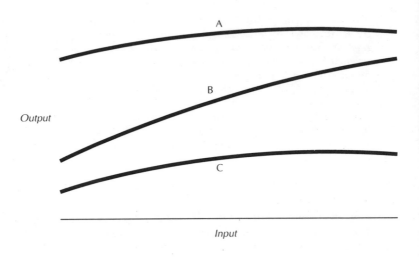

Figure 4 *A generalized response function depicting output-input for three
soils with varying levels of antecedent fertility*

(Lal, Eckert, Fausey and Edwards, 1990, 206)

Nutrient build-up has become a criticized aspect of the 'high
input' farming systems. Buttel *et al* (1990:516) cite US studies
suggesting both that 'traditional petrochemical-based production systems are reaching a point of diminishing return' and that
problems of pollution are increasing: 'Enforcement of pollution
standards in agriculture to the degree that standards are enforced
with regard to industrial corporations and municipalities would
have a devastating effect on agriculture in many regions of the
country.' Buttel *et al* (1990:518) remind us that agricultural
innovations have not always divided farmers from other interest
groups in society. When the first innovation in the area of
biotechnology came on the market in the form of growth
hormones for cattle, there was considerable resistance from
producers '... particularly in some of the major family farming
dairy states' as well as from environmental and public interest
groups. It was clear that with productivity increases those who
could afford the technology would become stronger while others
would be forced out of business.
    The principal beneficiary of various of the modern, 'high
input' innovations in livestock production has been industrialized agriculture. Family operations have frequently been

subordinated to the role of contract producer, receiving the animals, feed and management from processing firms or other related industry, providing buildings and labor, and being paid a fixed amount per pound produced. In this system it is the producer who has to suffer the consequences of animal loss due to disease, climatic problems and other risk factors that arise.

In regions where family agriculture has remained diversified or returned to diversification, however, there persists the feeling cited above, of a 'return' to an older system. Investments made in specialized machinery and equipment as well as training are questioned as to whether they're really worth the sacrifice. But the modern diversification can be based on specialized resources applied to new income-generating activities. Pumps installed for crop irrigation may be useful for fish-raising operations. Integration of poultry-beef or dairy, or hog-fish activities are possibilities in many countries. Research showing that most of the nutrients consumed by livestock are returned via the manure suggests the importance of more nutrient recycling and, as part of that, a return of manure to crop fields. In research conducted in the US State of Ohio, Barrett *et al* (1990:630) found that

> ... *less crop diversity can slow soil development.... The result is soil degradation that affects landscape-level processes. For example, such degradation reduces soil moisture-holding capacity, which allows greater leaching and surface runoff of soil, nutrients and pesticides. Less diversity may also result in greater crop losses to pests [insects and weeds, and diseases] in which the success of regional [field to field] dispersal of pests is enhanced, the rate of adaptation to crops by pests is accelerated, and the associated resistance of crops to pests is reduced.*

They lament the use of grain-grain rotations which have nearly eliminated hay and forage production in the state.

Some specialized linkages established by individuals for purchase and marketing of products and decision-making may no longer be sound, with diversification. This is where cooperative relations can become of renewed importance. A group of producers of similar products can actually be a more stable unit for purchasing and commercialization than independent individuals. In any case, diversification as a concept is based on agricultural organization developed over generations by family producers. As it is reexamined and found to be of possible future utility we are only reaffirming the advantage that this system holds, even in the face of our changing times.

Those who have studied forestry have suggested increases in tree planting for feed, fruit, fuel and lumber and as a component of diversification and sustainability processes. Trees can capture nutrients that have leached to subsoil levels and assist in the maintenance of the water table. Shade is important in subtropical and tropical countries for livestock and some crops (cocoa, tea and some varieties of coffee). An interesting example of international cooperation in tree crop research occurred involving the leguminous Leucena tree mentioned in the last chapter. This fast growing tree has the capacity to fix nitrogen in the soil, and is an efficient source of palatable forage and firewood. The University of Hawaii has worked together with several other countries in the promotion of Leucena. Experiments in Australia, however, reported a toxicity in both ruminants and non-ruminants when the small branches and leaves were fed in high concentrations. Researchers in India, on the other hand, reported no problems with toxicity. Scientists began intensive analyses and discovered a strain of bacteria which break down the toxin to a harmless substance and occur naturally in the rumen of animals in Hawaii, India, Philippines, Indonesia and several other countries, but were absent in Australia, Papua New Guinea and parts of Africa. The bacteria were introduced to Australia and are resolving the toxicity problem.

Research in biotechnology as a potential booster of production has gone far beyond the introduction of useful bacteria from one country to another. Genetic modifications have changed the characteristics of plants and animals to increase their usefulness. This research has not been cheap. The result is that biotechnology represents another series of what we have called 'high input innovations', of the category that increases risk and generally contributes to the present gamut of problems for most family producers. The individuals who adopt these innovations will be those with high levels of resources. In many countries these new techniques and products will remain 'exotic', out of reach because of cost and often not adapted to conditions different from the locale of the original research. For these reasons, especially in developing countries, biotechnological innovation is chiefly of interest to the capitalized firms such as those of agro-industry (eg feed companies and fertilizer plants) who are also involved in agricultural production. They are the ones who are able to purchase costly inputs for the more specialized, less rustic and potentially risky hybrid crops and animals. Recent losses due to disease problems in imported hybrid hogs from Europe cost a Brazilian agro-industrial innovator many thousands of dollars.

Theoretically it would be possible for biotechnology as well as other agricultural research to follow guidelines of ecological soundness and sustainability. But much research, by examining specific problems in an isolated context, ends up addressing some problems only to create others. Luna and House (1990:169), have listed some results that demonstrate this point:

- stimulation of aphid outbreaks in cole crops through the use of nitrogen fertilizers;
- insecticides increasing weed populations by killing the natural enemies of the weeds;
- use of carbofuran insecticide increasing the growth of crab-grass and other grassy weeds;
- fungicides killing soil fungi that exert natural control over nematode populations;
- insecticides and fungicides reducing earthworm populations, hence lowering soil fertility and water infiltration rates.

This is not to say that a program of sustainable agriculture would be better off attempting to eliminate innovations. New methods, biotechnological modifications and structural change are needed to meet evolving demand and merit attentive support, along with thorough investigation of all of their effects.

If the philosophical orientation that guides research were moved from profitability toward the preservation end of the continuum, however, the end product would be more useful to more farmers. Plant breeders could produce species that are less demanding of soil nutrients, plants able to tolerate pests without specialized breeding for pest resistance, plants that retain a broad range of genetic diversity for tolerance of varying climatic conditions with increased productivity. Animals could be bred for more efficient production on lower quality feed, especially forages, so as to reduce competition with humans for grain. But these are research decisions that are made on the basis of vested interests. They will only change as those with alternative interests organize and articulate proposals are put forth. Efforts are already being made to develop pesticides that are effective during a shorter duration, for application targeted at the specific time of the threat to the crop.

Energy use

Another aspect for consideration in the formulation of sustainable agriculture is the way that we use energy. In our discussion of Figure 3, we suggested that 'no till' and 'plow till' methods are principally employed for economic reasons. Conventional wisdom dictates that reduced tillage cuts down on the energy requirements of crop production. The research of Lal and colleagues affirms that, in the US, primary and secondary tillage operations consume 10 to 12 trillion K calories per year. Since a K calorie is a thousand calorie unit, this is a tremendous amount of energy. Then comes the surprise: 'In comparison, energy used in production of fertilizers is about 160 to 170 trillion K calories. Thus a reduction in frequency and intensity of tillage would reduce a small proportion of energy input' (Lal *et al*, 1990:207). Energy savings are to be had by reducing tillage, but methods such as the planting of legumes for nitrogen fixation, that diminish fertilizer use, will be vastly more effective. Edwards (1990) has reported that between 1970 and 1978, US farmers used 50 per cent more energy to produce 30 per cent more crops. His calculations work out to a requirement of three calories for every calorie of food produced. (Processing and distribution took another seven calories for every calorie of food.)

This is certainly not to say that conservation tillage is without merit. Table 1 presents data from field trials in Nigeria for a comparison of 'plow and no till' systems. Runoff water is nearly six times as much on plowed ground and tons of soil lost to erosion, even more.

Smaller, lighter machinery can be used as an intermediate solution for reducing the damaging effects of tillage without

Table 1 *Effects of a no-till system on runoff and soil erosion from corn for a tropical soil in western Nigeria*

Slope %	Runoff (mm) No-till	Plowed	Soil Erosion (t/ha) No-till	Plowed
1	11.4	55.1	0.0	1.2
5	11.8	158.7	0.2	8.2
10	20.3	52.4	0.1	4.4
15	21.0	89.9	0.1	23.6
Mean	16.1	89.0	0.1	9.4

Source: Lal *et al*, 1990, p 211

abrupt changes in cropping systems. Not only is soil compaction reduced but less energy is consumed for operations and the cost of the machinery itself is more within family budget levels. Recent low horsepower models of tractors that have come on the market in several countries indicate the sensitivity of the farm equipment industry to knowledgeable future demand.

Sustainable agriculture in many countries will be based on the present, unmechanized systems. Tiny Chinese farms don't permit efficient use of machinery as it has been developed in the western countries. The slash and burn agriculture of tropical countries, where fields are full of stumps, also doesn't present conditions appropriate for western style mechanization. Methods that have developed in areas of rigid tradition resist change, even if land conditions were suitable to permit acquisition of machinery. As mentioned in the last chapter, Latin American farmers debated with extension agents for twenty years over the advantages of planting beans under corn. In Brazil the extensionists were against it as an old fashioned idea, prohibiting mechanization and reducing yields. Now, with new research findings, the extension service has agreed that for best land use, and because of the symbiotic characteristics of the relationship, the two should be planted together. Actually in terms of maintaining a ground cover and as a method for the maximization of the use of solar energy, the 'succotash solution' is superior to monoculture. Where it is used, the farmers are not likely to become mechanized and have never doubted that it was the best system for them.

There are other specific characteristics of some regions that affect sustainability. The planting of legumes in the tropics and sub-tropics is a possibility for nitrogen fixation. Unfortunately they cannot compete in most places with the aggressive grasses, making it nearly impossible to maintain a grass-legume pasture for any extended time.

Swine production, on the other hand, represents a problem for sustainability in areas of population growth. Hogs and people eat the same things. In terms of feed conversion, of course, they're not as efficient as hybrid broilers. The fact that hogs are still raised in the poorest regions is explained by the fact that lard is the chief energy component of the diet and by their use of food scraps. Rabbits and goats, possible alternatives, are also able to make use of food scraps and crop by-products as well as low quality forages, thus furnishing meat without competing with humans. Although these animals fulfill, at least partially, the needs for protein, there is still the requirement for an inexpensive source of energy which, among populations involved in much physical labor, is of extreme importance.

CONCLUSION

We have looked at the question of sustainability and it appears to constitute an objective that must be considered for the welfare of future generations. The importance of reducing soil erosion and nutrient run-off has been well documented by the conservationists. These are coupled with the necessity to reduce surface and ground water pollution from agricultural chemicals and sediment. At the same time there is a need for increases in production and productivity to meet the expanding demand. For any of these needs to receive serious consideration from farmers there must exist the assurance of adequate and stable income.

How can agriculture be organized so that producers will invest in these goals? When there is need for short-term resources, sustainability will generally be relegated to a lower priority. This may be the case of families working to pay off debt or producers on rented lands or managers of industrialized agricultural enterprises. And these three roles may all be played by the same individual. The point is that short-term gain is not consistent with long-term survival.

Who is interested in long-term survival? Salaried workers see survival in terms of weekly paychecks and retirement benefits. Managers seek advancement through increasing firm profits. Owner shareholders seek returns on their investments to assure their survival. *The possibility of implementing sustainability in production systems as we know them is better in situations in which the producer is also the manager and investor.* Debt management will be integrated into other objectives, which may mean sacrificing a 'grow or die' ideology. Only those who are concerned with long-term production capacities of resources will be interested in sustainability. Are we describing family agriculture? There is, of course, a need for well researched and planned programs to meet the needs of sustainable production. In the real world the dialectical recognition of relations and change are known to be inescapable. With this realization the family producer may well be the key to survival as the producer of food and fiber for today's needs and the one who has the vision to protect resources for the future of our children.

6

Family Agriculture and Family Values

VALUES FOR WHAT?

Values are shared expressions of what we believe. While we often think in individual or personal terms, social scientists have defined values as characteristics of groups and populations. Individual *attitudes* express one's reflection (acceptance, modification, rejection) of these values. For shared beliefs to be promoted in society it is necessary that some procedures be established. Societies around the world develop methods consistent with their cultures for the formulation and promotion of values. Remember that, in accord with our innovative logic, values are *in process*. They are continually being evaluated and modified by all of us. In ancient times stories and songs were important media for value transmission. Even today, in West Africa and other rural areas visitors are asked to recite 'proverbs.' These little stories with strong moral implications reinforce value orientations. With modern forms of communication – the written word – it became less difficult to foster values in a population and recommend them to other societies. Literature, religious dogma, histories and other writings have been used in this process.

As societies develop, institutions are established to cultivate values. Religious education and community organizations as well as the commercial infrastructure serve as vehicles which, together with their specialized functions, endorse social values. Mass communication becomes important since huge numbers of people can be reached.

But values cannot be found under rocks. Where do they come from? Values are generated from the attitudes of individuals and groups who, through their persuasion and their power, are able to implant their ideas. Of course these opinions must be acceptable to those who are to adopt and reflect them in their personal lives. Over time some 'universal values' establish rules for social relationships.

With mass communications we are suddenly bombarded with competing value suggestions. Issues arise which polarize segments of the population as they seek acceptance of their attitudes among the general public. Arguments are often couched in terms of right and wrong even for subjects of controversy. It is fairly obvious that our attitudes are based on what we perceive is *best for us*. But when what is best for us involves other people, we must convince them that it is best for them, too. On some subjects, such as who should pay taxes, this can get complicated. On the other hand, if we allow the values of society to weaken in favor of our individual attitudes, our common future becomes a subject of question.

As individuals interested in our own well-being, we are vulnerable to commercial appeals for the 'latest' material items on the market. We are more easily convinced that it is best to 'go our own ways.' With urbanization our 'own ways' become separated during much of the time and peer pressure overcomes the influences of traditional family ties. So we must contemplate, values for what? Do we need commonly accepted norms of what we believe? Is it important to be able to express what we stand for as family members, community residents and, yes, family farmers? Is it necessary to have an ideological 'shield' to protect us from the glitter and boom of immediate gratification by way of credit card purchases? Who are we?

A CONTEMPORARY PERSPECTIVE OF THE FAMILY

Thinking of family values brings to mind ideas like respect, honesty, loyalty, the 'Golden Rule' and the importance of work. Individualism has broadened these to include topics such as security, accumulation, recognition, opportunities for service and some others that may be more pertinent for discussion in relation to family agriculture.

Some recent rhetoric, principally in the developed countries, has centered on the composition of the modern family. There are 'joined' families and 'broken' families and other adjectives applied as social forces mold contemporary social action. This issue, for family agriculturalists, is not so complicated. *The family is composed of those who work and play together*. Rural communities abound in stories of families who have lost, for one reason or another, members and have adjusted their activities to survive. Sometimes outside individuals have been accepted into the activities of family agriculture and are soon considered 'like family to us.' My suggestion is that importance does not lie so much in blood relations as in common activity. We need time together to influence, to help, to be helped and to appreciate one another. Accepting this definition we have flexibility to recognize varying combinations of age and sex groupings as families. People working, learning and enjoying life together in pursuit of their own objectives reduce the importance of the arguments of politicians who purport to define what a *real* family is.

This is not to say that those in family agriculture are not involved in divorce or drugs or any other social problems of the times. Rather it is that these families have a structure which can help to cope with undesirable situations. Changes in the family have occurred most in countries where there have been other forms of rapid change. As living standards have risen in the developed countries there has been innovation that reduces interdependence. No longer is there a need for one family member to remain in the home, occupied with cooking and cleaning. These innovations cost more than traditional appliances but as the house-bound family member has been freed it is possible to take on additional employment, to cover new payments.

Families with multiple wage earners often need to renegotiate household chores. When farming is part of the employment there are even more chores and more possibilities for family stress. This was referred to in Chapter 2, in the discussion of cooperative work groups in France. In the future we may look back upon the present time as a transitional period that required family adjustments. But for us, here and now, the problems are very real and, for some, overwhelming. These problems may not go away soon, but they can be absorbed into something larger and more important to those involved.

THE ROLE OF VALUES IN FAMILY AGRICULTURE

The value of work

Salaried employment, an activity that has come to be accepted as 'normal' in developed countries, is not so naturally assumed in other parts of the world. There, the idea of 'working out' or 'working for others' is considered undesirable and consented to only as necessary. Increased opportunities in industrial sectors and individual interest in higher standards of living, however, have lured families away from the land and into the cities. Many soon realize the disadvantages of salaried labor, but access to land is difficult, the costs of farming have increased rapidly and the road to the city has generally been a one way trip.

Social values in all countries influence individual levels of satisfaction. As long as urban jobs are plentiful and happiness can be found in the personal and material sense, efforts have concentrated on keeping up with instilled ambition. This is increasingly no longer the case. A recent article in *American Demographics* (Braus, 1992) reports that US employers are having a hard time keeping workers. The employees interviewed sought a more balanced life, and had less tolerance for what they considered 'dead-end' or 'busy work' types of activity. In response to questioning, the following objectives, among others, were voiced:

● more interesting jobs
● more job flexibility
● time for family and friends
● being able to work independently
● a chance to get ahead
● recognition from co-workers

Workers feel trapped. They sense a gap between their employment activities and the values that they seek to reflect. Separated from family and friends during most of the time, they must rely on salary levels for justification and satisfaction. Ironically, problems in the national economy help to keep workers more firmly in their jobs. The burden of monthly payments and the awareness that others are losing their positions make people swallow their dissatisfaction and reluctantly continue on the job. When the economy improves they hope to be able to change to something better.

Some countries have attempted to raise worker satisfaction by reducing the number of hours on the job. The Braus study provides an indication that this would not currently be an effective solution in the US. Those interviewed declared that they would be willing to work an *increased* number of hours per week for a job that they liked more. Also leisure time, in the absence of planned activities, resources and the time of family members, can become boring. Braus cites a Gallup poll of US workers conducted in 1955 that reported 38 per cent agreeing with the statement that they enjoyed their time at work more than the time off.

When this question was repeated with a similar sample in 1991, only 18 per cent agreed.

Security

A significant value in any culture is protection from need. Security becomes especially important when there are imbalances in societies that threaten well-being from a physical, economic or even spiritual perspective. A contemporary example in the developed countries is the incredible increases in health costs and the expenses of long-term health care. It is true that we are living longer and enjoying relatively good health for more years, but we might ask, for what? Speaking recently to groups of retired Americans I have been surprised by the number of expressions of nostalgia that are voiced for 'the farm life.' It appears that these people would be happier if they had more independence and more social contact, perhaps even in exchange for a few extra years of life. The possibility of saving the amount needed for health security in retirement, or paying the elevated health insurance premiums is, no doubt, difficult or impossible for most of the population. On the other hand, a living situation in which an extra bedroom can be built on and more garden planted to accommodate aging parents may be a more practical alternative for those who have this possibility.

Economic constraints have forced European families to live together in extended families for generations. Parental generosity has been largely responsible for the building of a secure life situation for young families. This has been reduced on an international scale as the individualism discussed above has become more of an accepted social value. There are positive aspects of individualism in growing societies where risk taking can yield benefits and high mobility supplies workers with appropriate skills to the labor market, even if they must move across the country. However, high levels of risk taking and geographic mobility in search of employment are characteristics of young societies. With maturity, establishing roots takes on more importance. Familiarity with neighbors and colleagues and even physical surroundings becomes more significant. Family agriculturalists may not recognize the increasing importance of these aspects in society because the changes that have created so much isolation and disturbance in the general society have been less drastic in rural areas.

From a cultural or spiritual point of view, contact across the generations to provide assurance of who we are is still a major factor in most countries. Remembrances of family historical events, land purchases, the processes of community construction and other significant happenings become stories told by grandparents and build pride among us just as do the trophies on the piano, or in the front hall of the local high school.

Recognition

The recognition of ability and contribution in work was cited as being 'important' by 62 per cent of the workers interviewed in the *American Demographics* article (Braus, 1992). Only 24 per cent of them, however, were satisfied with the recognition they received in their current jobs. Routine industrial occupations in which workers are interchangeable and the industrial product takes priority over those who produce it have limited possibilities for worker recognition. The Japanese apparently have developed mechanisms to increase the recognition of urban workers, although the stability and longevity of employer-employee relations in Japan are characteristics not found in most other societies.

And what of the recognition received by those in family agriculture? The times of clean fence rows and well maintained buildings have passed in many areas. But county and state fairs are still notable events. US farm related programs and organizations from the 4H clubs and Future Farmers of America to the Grange and the Farmers' Union have provided leadership training and experience that have strengthened agriculture as well as other sectors of American society.

Recognition is also important from a personal standpoint. When we have a situation in which we can exercise initiative and observe the results, we feel better about ourselves. The possibility of using our own ingenuity and dedication and receiving the results as our reward is a tremendous stimulus in recognition of our effort and ability.

Opportunity for service

The exchange of items such as food, breeding stock and work in rural areas binds the population together and provides lines of communication and customs that make it possible to respond when there is need. It is ironic that rural people, who are more physically isolated than their urban cousins, often have more meaningful contact with neighbors and friends.

An opportunity that we traditionally enjoy is the passing on of our knowledge, skills and values to the next generation. Needless to say there are limited possibilities for this to occur in the urban sector, although some effort has been made in urban areas to establish mentoring programs, to associate successful adults with children and adolescents in need of attention. These and similar formalized relationships are constrained by the isolation of individuals, even as they are concentrated geographically and socially, into categories of rich and poor, black and white, urban and suburban. Gifted youngsters are singled out for advanced educational programs while those in need of 'special education' are separated into schools where they risk being abandoned to teachers unable to find positions in schools that they would consider more desirable. Since the need for special education is related to poverty, those involved can do little to change the situation. In rural areas people of differing racial backgrounds and socioeconomic levels are still more closely integrated. Students with more, or less, interest and ability are still in the same classrooms with peers who serve as role models in value sharing and thus reduce the risk of drifting off.

Creation and recreation

An invigorating and exciting part of our lives involves the planning and expectations as well as the surprises of new life. Here too, farm life has some special characteristics. From the hen who steals a nest and reappears with a dozen chicks to the construction of a uniform and productive dairy herd, we garner great pleasure while learning the values of responsibility and persistence. These values are complemented by patience and self control in raising livestock. And they serve us well as we deal with the situations of life. Although most of us spend more of our time relating to people, the values learned in livestock raising transcend these experiences to prepare us for effective social relations.

The recreation of farm people has traditionally involved activities similar to their mode of life. Hunting and fishing and social events that provide face to face meetings for conversation, whether they are ostensibly for sewing or organizational planning or children's activities, are consistent with typical on-farm production undertakings. Visiting is still practised in many countries as well as more isolated regions of developed nations. Much leisure time is spent accompanying children's and youth activities: sports teams, drama, arts and crafts. In some regions the regular market days or seasonal markets provide times of great anticipation and enjoyment. In all of these examples people are joined together and topics of discussion facilitate the sharing, questioning, maintaining and changing of our value heritage.

Stress

When life and work situations contain aspects over which we have little or no control, and which are in opposition to the values that we espouse or our own personal attitudes, we experience stress. A few decades ago stress was considered an urban problem. The city life, especially for those who had migrated from rural areas, was different and likely to be considered unpleasant. The choice was not to work more or work less, depending upon need, but to work or go without.

More recently stress has been recognized on the farm. The US agricultural crisis of the 1980s was serious enough to distress nearly the entire farm population. Pressure to 'get big or get out,' loan payments, product prices, land values, taxes, all added up to a catastrophe that drove many good farmers from the land and severely castigated those who were able to withstand the pressure. It is necessary to examine this stress more closely to discriminate between urban salaried job stress and that experienced by farmers.

In the Braus study cited earlier, 58 per cent of the workers interviewed considered activities to reduce job stress 'very important' to them. Only 18 per cent, however, were satisfied with their current jobs in this respect. While there is some possibility for negotiation, individually or through unions, the text of the study made it clear that most workers thought less of changing their jobs to meet their personal interests, than of changing jobs. Of course in a period of economic recession the opportunities for changing jobs are very limited, which in itself adds to the stress.

Farm stress has characteristics of a different nature. While there has been the possibility of losing everything by some, and this could include farms that have been in families for generations, for most the stress has been more associated with the responsibility of managing large quantities of capital under conditions of extreme uncertainty. This is particularly difficult when the situation is structured so that a disproportionately large part of the risk falls upon the farmer.

The measures that can effectively be taken by urban workers, to be sure, are different from those in agriculture. For farmers, who are less mobile and can't change jobs so easily, values associated with cooperation, negotiation and political organization have become essential for survival. In this regard stress can be compared to pain within the body of an individual. It is a sign that something is wrong. And like pain it should not be ignored. While the ideas of unionizing and negotiation are abominable to many in agriculture, they are absolutely necessary. The farmer who just wants a fair price for his honestly produced grain and livestock is one of those nice people who will finish last.

Values associated with agriculture must be modernized to defend those who hold them and create and maintain respect for the agricultural sector among the population in general. Surprisingly this may not mean just producing more for less – the *quality* of production is increasingly relevant. New ideas for product distribution, rural-urban integration and other innovations require the formulation and support of legitimizing values.

Sex education

What could be an afterthought in a discussion of values may be of considerable significance in the family. Adequate knowledge of sexual reproduction and the ability to successfully breed livestock, and plants too in some cases, are critical for those in agriculture. Through knowledge and practical experience the stigma is largely removed from the topic of sex and related human social problems can be effectively confronted. Subjects that would rarely be brought up in other environments are routinely debated in farm families, putting a positive aspect on the associated values. Sterility, as in mules, and hybrid vigor, as in corn or hogs, are commonly discussed along with triple crosses and implanted embryos. The farm child grows up aware of the necessity of males and females for reproduction and the roles of each. Learning in a situation where there is no large group interaction among those who lack knowledge avoids much of the giggles and embarrassment in relating to the person who seeks to impart the knowledge. In the informal and familiar situation of the barnyard the child senses less need to mask his or her innocence and is not so ashamed of curiosity.

Sex education is handled in keeping with the cultural norms regarding the division of responsibility in the family. Men in West Africa have been known to look away when coming upon a cow giving birth and go directly to inform their wives who 'know how to take care of that.' Norms change over time and even in more open societies like the US where sex recently seems to be one of the most popular topics on television, adults who were not raised with such open discussion sometimes have difficulty educating children. The natural environment of farm life, however, relieves the tension. Children learn in ways that are still not extremely different from the experiences of their parents. In the case of families with missing parents, the farm environment will actually assist youngsters in this learning process, in compensation for the more complete explanation that might have been provided.

CONCLUSION

We can conclude that family agriculture as a production structure is greatly benefited by the characteristics of the family itself. As families change, both in composition and in relations among members, however, stress is created within individuals. Where there has been more change, there is more stress. While all families in such situations may experience these problems, the structure which holds them together and the activities which intensify their relations with each other make family operations part of the solutions rather than part of the problems. As values change, farm families need to be remembered as models for others in society.

From our dialectical perspective we recognize that the family is being transformed, and it would be ignorance to pursue an historical form of social organization in an idealized manner. Families in agriculture are still part of the society in which they live and subject to overall changes in values. As the idea that divorce, for example, is better than a problematic marriage, has become more accepted, the definition of 'problematic' has changed. Irritations that were previously taken into stride have been given increased importance and there is less chance of working out the problems. This causes friction in any activities that are dependent upon joint investment, joint labor and joint decision making. Any breakdown of the partnership involves a complex division of property that handicaps both parties in terms of the possibility of continuing in agriculture. Land based activity is especially hard to divide and creates more of a problem to continue in altered form. Still, we must recognize that marriage partners on the farm have high levels of face to face contact and thus a structure for negotiation. Perhaps the best that can be hoped for is an understanding that the family unit, however it may be formulated, is a structure so precious in society that it behooves us to subordinate other values and concerns to its perpetuation. And we must promote this value as an objective on which future generations of family agriculture can be constructed.

7

Food Security

Food production in the developed countries is being transformed from family enterprise to big business. As we become more convinced of the theory that 'time is money' it is difficult to accrue the motivation to plant and raise and kill and prepare and buy and share food items as has been the practice of past generations and is still the case in the developing parts of the world. Considering the hours, a factory job often looks better. Our implicit belief is that we are 'progressing,' that our standard of living is higher. The health, environmental and social questions that are also progressing, may be seen as the kind of problems that other countries wish they had. But they are still worrisome....

Discussions of food security bring to mind the panic of scarcity. The mass media reports tug at our hearts with stories of emaciated children walking miles in search of some hope for survival only to die before help arrives. The issue of adequate high quality food on an international basis, however, is very complex. There are good reasons why families will continue to have an essential role in the process. These reasons add to the formulation of our response to the question raised in the introduction: Why is family agriculture important?

With the enormous increases in population since World War 2 the immediate question is, where will the food come from? Why is the population growing so rapidly? Three basic factors explain reductions in infant mortality and longer life expectancy in today's world:

1. the discovery of antibiotics;
2. improvements in hygiene; and
3. the development of vaccines.

In the developed countries these factors have been tempered by reduced birth rates. Although other parts of the world have also shown some signs of lowering numbers of births, higher survival rates boost growth indices very quickly (Figure 5). Farm families in these areas still see children as assets to the family labor supply. Urban families are often more influenced by religious orientations than by family planning clinics. The perception is that God will provide.

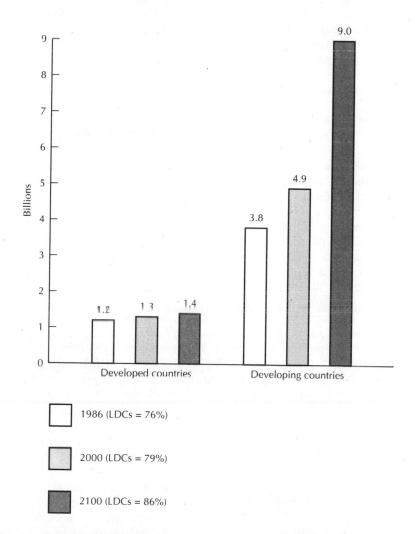

Figure 5 *World population distribution – 1986, 2000 and 2100*
(Taken from Brady, NC, 1990)

This is not to say that the commonly accepted ideas of what has been called 'planet crowding' are necessarily the most appropriate for our inquiry into food security. Some time ago the Brazilian governor of a large northern state dispelled, in his words, the myth of overpopulation, declaring that he could accommodate the entire national population in his state, providing several acres for each family. The rest of the country would be empty. His point was that hunger is a problem of too little economic growth rather than too many people. While these examples appear to oversimplify the situation, they demonstrate our need for a 'handle,' a focus to examine the importance of food and how it is related to family agriculture.

AVAILABILITY, ACCESS AND ENTITLEMENT

The concept of security suggests considerations of supply and availability. Economists examine supply in relation to demand with the objective of avoiding large fluctuations in price. Agronomists plan for the reliability of the food supply and the sustainability of production processes, as we have seen in previous chapters. Researchers who have specialized in resource planning over the last twenty years, however, have emphasized that food problems are more associated with distribution than with supply. We must thus recognize the importance of *access* to food in addition to its availability.

At the same time that farmers in the developed countries have been called to feed the world's masses, there seems to be no end to the news photos of starving African mothers cradling their withered and dying children. We can understand this suffering better through the idea of 'entitlement,' a concept used by Hartmut Schneider (1984:112) at the Development Centre of the Organization for Economic Co-operation and Development, headquartered in Paris. National laws or norms of what different categories of people in a society are entitled to vary dramatically. The developed countries are sharply focused on the rights of the aged and of national racial and ethnic groups. In the developing countries the conflict is more likely to involve differences of race, religion and geographical origin. Entitlements take on increasing importance with social and economic development, and are often threatened during periods of scarcity. From this point of view one can more easily understand the TV footage of healthy looking young men unloading cargo planes in countries where we imagine everyone to be hungry. A citation from *The New State of the World Atlas* (Kidron and Segal, 1991:129) reinforces the point:

> *People seldom go hungry because there is not enough food. More often it is because they cannot command an adequate share of what food exists, or the material and cultural resources to make good nutritional use of what they might find before them.... The world is capable of feeding decently all its inhabitants. That it is conspicuously not doing so at present is the product not of necessity but of choice.*

Starvation is a political instrument for genocide, to promote ethnic or religious purification as well as the elimination or weakening of any category in the population considered to be undesirable by its leaders. An example of this was reported in *Newsweek* magazine on 12 October 1992: Sudan, in Northeast Africa, is a country composed of Africans in the South and Arabs in the North. These racial differences are further emphasized by contrasting religions: the Arabs are Muslim while the Africans are mostly Christian or animist. Armed conflict is frequent between the two groups. Casualties, which are more often African, are generally explained by the Arab controlled government to be victims of starvation. The article, however, reports that the Arab government of Sudan had just donated 1000 tons of food and medicine and 500,000 dollars 'to help suffering people in Muslim Somalia.'

There is less stigma and social condemnation attached to starvation than, for example, to the methods applied by Hitler in purging the European population of its Jewish and other minority residents. Somehow starvation is associated with providence and helplessness. As resources dwindle, the need for reductions in the population becomes a method to make the food go farther. In a kind of social cannibalism, politically weakened segments are not physically consumed but are eliminated before they can reduce available stocks of food. To be sure, the suffering is very real. But one thing is clear: we will not eliminate starvation in the world by producing more food to export from the developed countries.

WHO PRODUCES FOOD

If we classify agricultural production into food crops for consumption and local markets, and cash crops for national and international markets, it is the latter that offers more possibility for profit in most of the world. Cattle, sugar cane, coffee, cocoa, tea, herbs and others have traditionally been raised and grown for national and international markets. This distinction has become somewhat blurred in the developed countries with the elaboration of industrial uses, for example, of corn and soybeans.

The Brazilian Agricultural Extension Service published a study a few years ago showing that between 80 and 90 per cent of food crops: rice, edible beans, cassava, and others, were produced by families. This family production is organized as was common in the developed countries not so long ago. Food is raised for the family and the excess is sold. In many cases they also produce cash crops for monetary income. The emergence of large, industrialized farms which can deliver regular quantities of standardized product, however, has cut into the market of families who were accustomed to taking a crate of eggs to town on Saturdays. Families have been forced to adapt modern ways, increasing the size of their operations and improving methods, or going out of business.

This transition has brought about some interesting adaptations, on the part of families, to the necessities of modern production. In the Zona da Mata region of southern Brazil small farmers recognized early the necessity of bank credit for the purchase of fertilizer to increase crop yields. After some disastrous experiences in which families lost their land due to an inability to repay loans, they adapted to the formal credit system. Request was made to the bank for a loan in the normal way. A plan was drawn up for application of the credit, usually for the production of crops. Then, when the first parcel of the loan was received, a calf or two would be purchased with part of the money. The remainder would be applied as prescribed by the loan agreement. In the event of crop loss due to weather or the uncertainty of market conditions the calves, then fat cattle, would be sold to pay off the bank. These families continue to produce food crops and, although without the bank's sanction of the procedure, are in good financial standing. This is just one of innumerous instances of family adjustments to modern agricultural production.

Economists speak of the inelasticity of food crop prices. We are unlikely to buy very many loaves of bread at one time, even if it's on sale for a fraction of the normal price. This is a factor in the decision of industrial farmers to produce food crops and has resulted, in the developed countries, in great quantities of research funding for new, more resistant, durable types of foods that can withstand mechanical handling and have a long shelf life. To the extent that such food crops have been developed, larger than family farms have increased their share of the market.

Still, the existence of a secure and reliable source of food is a necessity of all nations. In some countries increasing the production of food for the market is especially problematic. Locally adapted crops may not be responsive to chemical fertilizer. Imported seeds have been bred for the soil, weather and climatic conditions of temperate climates and generally don't do as well in tropical or semi-tropical areas. Local food crops, for example sorghum, millet and cassava, on the other hand, are considered low risk (high security) crops. They are less demanding of soil nutrients, produce even in dry years, and have fewer problems of disease and insect infestation. But harvests vary drastically. When an agricultural extension agent in Togo, West Africa, was questioned as to the limited success of a campaign to encourage the use of chemical fertilizer in his canton, he replied: 'It's hard to convince farmers to use a new product that may increase their harvest by up to 20 per cent when yields normally vary by more than 50 per cent, depending on the rains.'

No general, international guidelines are likely to be effective for increasing food access to the people most in need. Every country has its particular characteristics and traditional relations with other nations. In the poor countries food prices may not be sufficiently remunerative to stimulate increased production. Even in areas of rapid industrialization, as cited previously, it is in the interests of industrialists, who generally wield considerable power in the government, that food prices be kept low to maintain workers' living costs (and indirectly salaries) at minimum levels. The result is a perception by farmers that there is little opportunity in the production of food for local markets. This leads to growing interest in producing for export.

TRADITIONAL FOOD CROPS V CROPS FOR EXPORT

In 1987 Tom Barry reported that the US Agency for International Development and the World Bank were urging farmers in Guatemala to change from the production of local food crops to cultivate Brussels sprouts, snow peas and broccoli. Since these foods have very low acceptance locally, they have no direct effect on local food security, but could provide export earnings; that is, if the international commodity prices remain favorable to

Guatemalan produce in relation to that from other possible sources. Even in this case there is no assurance that income from exports would be used for food, or would be available to those in need of food. Barry saw more opportunity for US investment in the Guatemalan situation than for the local farmers (who would be the principal recipients of increased risk), or for local food security.

Information from other Central American countries also demonstrate that the production of food itself may threaten national food security. Whiteford reports that at the same time that Costa Rica became a major exporter of beef to the US, 'There was a net decline of 26 per cent in local per capita beef consumption' (1991:133). In Mexico a massive investment in irrigation infrastructure over a period of 39 years has increased agricultural production but it has also resulted in areas of salinization and siltation: 'Already over one million acres of land that were formerly cultivated are no longer usable because of these ecological consequences' (Whiteford and Ferguson, 1991:15).

The result of these and other problems is that individual food consumption has been reduced. In Honduras, Susan Stonich (1991:45) found that during the last generation the production of staple foods '– maize, sorghum and beans – declined [on a per capita basis] by 24 per cent, 83 per cent, and 44 per cent respectively.'

THE PRINCIPLE OF COMPARATIVE ADVANTAGE

Agents of international development have long advocated that nations should produce that for which they have an advantage in terms of land, related resources, climate, labor costs and other factors, and trade with countries who produce needed goods. The Latin American countries have subscribed to this orientation for many years. But it has become apparent that the industrialized countries want principally raw or semi-processed materials from the developing countries, which are then processed, concentrating job opportunities in the countries that already have considerable wealth. A Liberian storekeeper complained that the German Mining Company takes African iron ore to Europe and 'They send back hammers that cost too much.' This also happens with agricultural products. Even when food produced in countries with a comparative advantage is donated to countries in need, there are problems. Algemiro

Brum, a native of the wheat producing region of Brazil, wrote in a book (1988:38) that has been widely read in his part of the world:

> *Beginning in the 1950s the North American Government began using wheat as a political weapon for various purposes. It was supplied under extremely favorable conditions to countries like Brazil, Colombia and others. In 1954 Brazil signed its first Wheat Agreement with the US and began importing increasing quantities of grain ... under conditions apparently quite advantageous: low prices, long term loans for repayment (up to 40 years) and low interest. As a consequence of the ease of importing US wheat there was a lack of concern for our own production, principally with respect to genetic research, a reduction of imports from Argentina and Uruguay, our traditional suppliers, and damage to the economies of each of these countries.*

As wheat processed into bread became less expensive and easier to serve than local staples – corn and rice products – the taste of most Brazilians evolved to 'salt bread' and the country has been set back, compared to other nations, in its ability to respond to the demand. Brazil and most importing countries would be seriously hesitant to enter new agreements to become dependent on US wheat after witnessing the grain embargo against the Soviet Union in 1980 and the later food embargo against Nicaragua during the Sandinista regime. Reliability is a chief factor in justifying national production of staple foods.

SALVATION THROUGH TECHNOLOGY

Much emphasis has been given, and is still being given, to technology as a promising solution to the need for more food. The 1992 *Ohio Farm Science Review*, an annual American exposition, featured a new six-row combine header for picking corn that was shown in color, on the front page of capital city newspapers. This mechanical technology has been joined by innovations in chemical and biological technology over the years. These innovations have driven up the price of getting into – or staying in – farming, at the same time that they have made it possible for one person to produce much more. This is, as was suggested before, exactly why much of the technology from developed countries is not appropriate in developing country situations. Expensive labor saving equipment in countries with extremely limited supplies of capital and high levels of unem-

ployment doesn't make sense. The obvious response from the
farm sector has been to reduce the population so that fewer
producers will have more capital to invest and more opportunity
for profit. This is a fixed form of logic. It would be more
appropriate to question, up to what size do farms need to be
increased and labor be removed, to attain economic feasibility?
This is to say, what scale of capital, land, and labor units are
most efficient? Vernon Ruttan, well known agricultural economist at the US University of Minnesota, has made notable
contributions in the evaluation of technological change in
agriculture. After intensive research in various countries he has
recently written his 'personal perspective' of the situation in
which he affirms (1988:50) that '... less developed countries are
operating in the region of constant returns to scale.' In other
words, increasing farm size will not increase economic efficiency
per acre regardless of the levels of technology that have been
adopted.

And if the developed countries are included in the analysis Dr
Ruttan (1988:55) claims:

> *I see nothing in the evidence presented in the recent rash of
> technology assessment studies that leads me to anticipate
> productivity gains over the next several decades comparable
> to the gains achieved since 1940.... We can expect a slowing
> of additional gains from advances in mechanical technology.
> It appears to me that the cost of saving an additional man-day
> by adding more horsepower per worker has largely played
> itself out in countries like the United States, Canada, and
> Australia.*

Hartmut Schneider studied data from research on agricultural
technology in 70 countries. His analysis (1984:103) speaks of
'scale neutrality' in agricultural production: 'It seems to be
generally recognized that the yield increasing technology can be
used at least as efficiently on small as on large farms.' But our
concern is not large versus small farms but family operations. He
continues: 'There is evidence that actual availability of certain
inputs, in particular fertilizers and control of irrigation water,
tends to be greater for large farmers. On the other hand, small
farmers tend to benefit from more motivated and cheaper labour
(family labour) than larger farmers.' The 'large farmers' are thus
not only those with more land, but what have been called 'larger
than family farms,' organized to use hired labor and with the
objective of producing profit.

ETHICS AND ECONOMICS

It has been estimated that between 500 million and one billion of the world's people are undernourished (World Resources Institute, 1992–93). This number appears to be reducing as a percentage of the total population, although it is still increasing in absolute numbers. According to the same source nearly all developing countries, with the exception of the African nations, have 'substantially increased' food production since 1970. The centrally planned economies of the Far East, China, Cambodia, North Korea, Mongolia and Vietnam have maintained production well ahead of population growth. The Near East and South America, however, barely managed to break even in the increases between population growth and increased food.

A debate that became classic of the thought of the times evolved in the 1970s. Environmentalists used the image of a spaceship to promote the idea that we are all in this together. Natural resources, therefore, should not be wasted, pollution must be controlled and population growth minimized in order to avoid, or at least delay, inevitable disaster. This proposal was countered by Hardin (1985:108–9) in whose vision the world situation could be more appropriately characterized by a lifeboat:

> *If we divide the world crudely into rich nations and poor nations, two thirds of them are desperately poor, and only one third comparatively rich, with the United States the wealthiest of all. Metaphorically each rich nation can be seen as a lifeboat full of comparatively rich people. In the ocean outside each lifeboat swim the poor of the world, who would like to get in, or at least to share some of the wealth. What should the lifeboat passengers do?*

His conclusion was uncharitable: 'Suppose we decide to preserve our small safety factor and admit no more to the lifeboat. Our survival is then possible, although we shall have to be constantly on guard against boarding parties.'

Looking back it isn't difficult to perceive that both sides of this debate presented the world situation as fixed, an over-simplification. They lack dialectical appreciation for dynamics and differences, for the comprehension of change. Every country has its rulers who largely determine the distribution of resources. It is not unusual in the countries of Central and Latin America, for example, for hunger to co-exist with the production of valuable export food crops. Achieving national potentials for local food production all over the world has frequently received lower priority than other political objectives. But the universal appeal, couched in ethical and moral terms, is that people are starving and this is the alarm that has been sounded to stimulate families to produce more food.

To set our discussion in its historical context, we can use as an example the situation that existed after World War 2. The United States was the dominant world power, with the dollar established as the world currency. Harriet Friedmann explains (1990:196).

> *This new international position put the American government in a unique position to find an international solution to domestic agricultural surpluses, a problem stemming from New Deal farm problems and exacerbated by the loss of European export markets through import substitution. After 1947 the Cold War isolated the trade of West and East, forcing the solution to be confined to the West. The solution involved two decades of export subsidies ('food aid') to Third World states, complementing the price subsidies to American wheat farmers.*

The Public Law, P L 480, was passed by the US Congress in 1954. Dan Morgan reports (1985:104) that

> *In the first years, an average of one bushel of wheat in four and one bushel of rice in five ended up going abroad with P L 480 financing. In 1959, a particularly bleak year for the grain trade, four out of five dollars' worth of wheat exports and nine out of ten dollars' worth of soybean oil exports were so financed.*

The program disposed of surpluses as it combated hunger. 'But when P L 480 first became law, a headline in the *Forbes* business magazine revealed the real power behind it: "Feeding the World's Hungry Millions: How It Will Mean Billions for U.S. Business"' (Hardin, 1985:111). The *Forbes* article referred to benefits to farmers and to manufacturers of farm machinery, fertilizers and pesticides, and to grain elevators and railroads and ports and shipping lines. All this required vast government bureaucracy and created vested interests in the permanence of the program.

Now, nearly 40 years later, another article from *Forbes* confirms part of what had been predicted. The 1991 list of the 400 largest private companies in the US, as reported in the Ohio *Columbus Dispatch* (25 November), cited the following:

Cargill, the Minneapolis-based commodities marketer, held onto the No.1 spot for the seventh straight year with estimated revenues of $49.1 billion.... Repeating as runner-up was Koch Industries, a Wichita-based oil and agriculture company, with revenues estimated at $19.25 billion. Staying at third was Continental Grain, a commodity trading and processing concern based in New York with $15 billion in revenues.

Part of that predicted came true. But not all ... US farmers, just recovering from a disastrous decade in the 1980s haven't seen that kind of wealth. At the same time that US farmers have been transformed into specialized producers of raw materials for an uncertain market, developing country farmers have been estranged from their traditional local and national markets by the importation of cheap grain. Families on both sides have been hurt.

Food exports and programs and subsidies have all been increasingly criticized by the general public. We have become somewhat numb to the ethics of furnishing food to starving people and are unaware of the ethics of manipulating a large part of one sector of the economy in favor of another. Sometimes the focus of the criticism has been on family agriculture as a relic of the past that can no longer be supported. This is largely a case of blaming the victim. It is not families who have caused imbalances in production. Grain gluts, as we spoke of in Sweden, and as have occurred throughout the developed countries, are the product of mechanical, chemical and biological innovations developed to increase production and avert supply problems of food stuffs. The contradictions that have arisen involve how to handle excess production in some nations, and how to avoid starvation in others.

International statistics demonstrate that food production in most parts of the world is on the increase. At the same time, the interest in donations is fading. John Mellor, an international agricultural economist, writes (1988:80): 'In most instances, neither the rulers nor the ruled of most developed countries feel strongly motivated to push for increased amounts of economic assistance or food aid to the poor countries of the world.'

PROSPECTS FOR EXPORTATION

Noting the present situation, what is the future of agricultural exportation from the developed nations? The countries that have resources to pay for imports are largely self-sufficient or committed to established trade relations for their food and feed needs. In the European Union, Swann credits the Common Agricultural Policy (1988:217): 'The CAP was also supposed to provide certainty of supplies. If certainty is to be equated with the achievement of greater degrees of self-sufficiency then the policy has indeed met its objective!'

Casting surpluses into the world market will only reduce prices, making it more difficult for developing countries to strengthen their agricultural sectors. Where there is need, mostly in Africa, and to a certain extent in Central and Latin America, there aren't the resources for market purchases. Still we hear the pleas: in Brazil – 'Plante que João Garante' (Plant, João [Batista Figueiredo, the last military president] will guarantee it); in the US – 'We can save the world!' And here lie some of the real ethical questions alluded to above. In the absence of an established policy, and without future contracts in hand, is it really ethical on the part of industry or government to encourage production for export? When this question was posed to an agricultural economist friend recently he responded with assurance:

> Just look at China's growing textile industry and you'll understand. With more exportation, salaried workers will have more resources and generate more demand. It only takes a small increase in the economic position of families to want to have chicken on the table more than once a week. With that many people there will be tremendous demand. . . .

We examined the writings of Harry Oshima earlier to understand that part of the world. To recap, this is what is said (1987:60) in his discussion related to food production:

> In the early stage of development, the demand for food by the expanding population is the most urgent, so that import substitution should start with the reduction of food imports for importing countries and the expansion of food exports for others – in both cases to obtain foreign exchange to purchase the powered spindles and looms and other machines for labor-intensive industries.

In other words, concentration on increasing national food production, as is presently occurring in China, will improve the economic situation and liberate capital for the importation of machinery.

The future of food exports in the world market seems

precarious. Regional relations are being formed to alleviate the problems of individual nations and create opportunity. The Japanese Emperor has gone to visit China, the crowned heads of Europe are meeting for the first time in many years. These symbolic meetings bring the countries together for the opportunities of increased cooperation. While developments may not produce great perspectives for profit, they may be quite positive for guarantees of food security and a more stable family agriculture on a global basis.

I mentioned Africa as being of special concern on the question of food security. Although there is great variation across the continent in terms of conditions for agricultural production, the ability of the population to produce sufficient food has declined seriously since 1970. Brown and Thomas have calculated that 140 million of the total population of 531 million people were fed exclusively on imported grain as early as 1984. This increased to 170 million or 32 per cent of the total population of the continent, fed with foreign grain in 1985. Brown and Thomas (1990:354) explain that:

> Even parts of Africa traditionally regarded as productive showed a decline. These lower yields are probably related to civil war and to environmental factors such as deforestation, over-grazing, soil erosion, prolonged drought, and generally inappropriate land use.

We have discussed cases of political and agricultural problems associated with food security. The plight of Africa is the most critical of examples. What can be done to break the continued dependence on imports and raise possibilities for self-sufficiency? Brown and Thomas, after studying the questions of increased population and depleted resources, suggest (1990:354–6) the following:

> ... reduced nutritional self-sufficiency ... leads to increased external debt and lower living standards. With such a large population and environmental deterioration undermining economic progress all across Africa, the only successful economic development strategy will be one that promotes and sustains the natural ecological systems on which the economy depends.

As in other areas that we have examined, the most promising alternative is 'African solutions for African problems.' Neither the most altruistic nor the most benevolent foreign entity can substitute the basic research and national development programs

that will improve the quality of African life through security of
its food supply.

BIO 'BEYOND BELIEF' TECHNOLOGY

Research in the area of biotechnology has yielded some
astounding new applications for traditional food and feed crops
as well as changing the characteristics of crops and livestock in
ways that diminish the primacy of nature in the production
process, reduce risk and raise possibilities for profit. It is not
surprising that, as a result, there have been notable increases of
corporate investment in agriculture. Harriet Friedmann
(1991:72) stated that 'Food is no longer simply something
produced by farmers and bought by consumers, but a profitable
product of capitalist enterprise, transnationally sourced,
processed and marketed.'

Capital inputs are purchased to increase production and
productivity and have produced admirable results. An article in
the Ohio *Columbus Dispatch* (20 February, 1992) reports an
example from an Iowa State University study that found total
costs of $38.20 per 100 pounds of pork produced by the most
efficient one-third of Iowa pork producers who raise hogs along
with grains and other activities, versus $34.14 for 'intensively
managed farms.' Pigs raised per sow per year were 15.01 for the
diversified producers versus 18 for the intensive units. A common
misgiving among traditional hog producers is that 'I can't
compete with that.' Of course the hybrid hogs used in many
specialized operations were never used to consume poor quality
feeds that would have been a loss, nor were they raised in any but
the optimum conditions that specialization can provide. The
traditional producer saves gilts from his own production for
breeding, depending on market trends of both grain and hogs,
and the needs of his own operation. Numbers of breeding
animals can, in this way, be more than doubled within a year.
Producers of genetically improved hogs, on the other hand, must
continually purchase breeding stock. The genetic vigor, which
increases efficiency in weight gains by the hogs, does not carry on
to the next generation. Dependence on a supplier of breeding
stock thus becomes permanent. Overhead costs of specialized
operations are also considerably more than are generally found
among traditional producers. This is not to say that the galva-
nized farrowing crates and intensive management, with sows
confined in an area in which they can take only a single step

forward or to the side, have solved the problems of swine production. The continual chewing by the animals on the pipe dividers is a sign of the stress that characterizes such operations. High overhead costs increase the difficulty of generating profit since the fat hogs, once off to market, sell for the same price per pound whether they are genetically hybridized, from a highly capitalized unit, or from a traditional, diversified, pork producer.

The inelasticity of demand for food products as already mentioned is another relevant factor. People aren't eating more pork because some of it is being produced under specialized conditions. Specialized producers can only increase profit if they increase their share of the market. To do this they must convince farmers who have hogs integrated in their diversified operations that they can't compete. There is actually nothing about hybrid hogs or feed supplements that cannot and will not be attained, with time, by family agriculture. In addition, the diversified producer who raises hogs has the advantage of cutting out middle profits when he uses his own grain for feeding. There can even be serious problems for specialized industrial concerns or farmers who become integrated with processing plants or other related industry. The high installation costs, cited above, responsibility for insurance, taxes and total risks of production are just a few. Payment is generally per pound of pork produced. Deaths or weight loss due to sickness are the problem of the producer. Disposing of manure, a common dilemma of concentrated livestock production, recalls our line fence neighbor who was put out of business by the townspeople two miles away because of the summer 'aromas'. He converted the buildings to machine sheds but much of the investment was lost.

Biotechnology has made it possible to substitute among crops to produce food, feed, fuel and raw materials for the chemical industry. Crop destination depends upon price and a series of considerations controlled by the big business that has become involved. An example is high fructose corn sweetener (HFCS), which is actually more expensive to produce in the US than cane sugar, but is a by-product of the corn wet milling industry that also produces gluten and ethanol, an alternative to gasoline. In Brazil this fuel (there manufactured from sugar cane) is currently being used to power more than a million cars. The corn gluten, an additional by-product, is used for feed. The destination of corn has thus been diversified among the markets of the by-products and depends not only on national markets but may be shipped in these various forms anywhere in the world.

A real surprise, however, comes from British food manufacturers who are producing an ingredient called myco-protein
using corn gluten as the nutrient substrate. Any other type of
grain starch can also be used in this process. The news media
reported a few years ago that a team of British scientists had
searched the world for a microscopic fungus that is edible and
would reproduce itself rapidly under conditions of continuous
fermentation, transforming low quality grain products into high
quality protein. They reported finding a suitable fungus just
twelve miles from their laboratory. Goodman (1991:47) reports
that it has been possible

> ... to produce a protein-rich ingredient, myco-protein, a
> micro-fungus whose textural properties can be used to imitate
> meat, poultry, or fish protein. The new product, with the trade
> name Quorn, is sold by Marlow Foods to food processors and
> is the main ingredient of a 'savoury pie' marketed by J.
> Sainsbury, Britain's largest food distributor, which is adver-
> tized as being rich in proteins and dietary fibre, low in calories
> and cholesterol free.

The substitution of industrially produced protein for animal
protein puts meat producers in direct competition with industry.
Related research reported by the same author predicts that milk
protein will some day be extracted from plant seeds. A frequently
used justification for these drastic changes is that land will be
liberated from feed grains and pasture, for the production of
food grains. Consistent with this line of thought, farmers will be
more and more treated as producers of raw materials, exploited,
as the developing nations have been, while the jobs and profits
flow to industry. But industry too cannot all be lumped together
as favoring such change. Researchers of food and pharmaceu-
tical corporations in Japan have also produced single cell protein
(SCP). Goodman cites a study attributing the failure of this
project to a campaign launched by the grain companies to create
consumer doubt as to the health implications of these proteins. A
reduction in livestock production would seriously disrupt the
grain trade, until it becomes linked into myco-protein production.

In addition to the new diversity of food, feed and industrial
output that is being, and will be, processed from traditional
agricultural products, the characteristics of these innovations
permit them to be stored for longer times and shipped longer
distances. As new marketing channels are established some

traditional products also benefit. Friedland cites the case of Greece entering the European Union as an example (1991:20). The new market for Greek raisins has had great effects on the production of American raisins and even on the US wine industry. Many other crop and livestock products all over the world are suddenly competing in this global market.

Perhaps the idea of eating fungus pie will be taken in its stride by the British, and soon by the world population. It hasn't been so long ago that canning meat was a 'dangerous' new idea in the developed countries. It quickly became a labor saver when families were pressed for time to shop or to prepare meals. The concept had possibilities for the developing countries as well, where refrigeration is often a problem and protein shortage is chronic. In spite of this, however, canned meat in those areas has not gained general acceptance. There are periodic rumors in Latin America that the imported cans contain horse or dog meat and in Africa it is suspected to be human! The new products of biotechnology will no doubt create some real excitement in such areas. . . .

All this extensive change may seem discouraging for family agriculture, but it doesn't need to be. Many of the same innovations that have attracted interest in agriculture by industry can be applied by families too. An advance that has great potential for rice production was announced by *Science News*, (17 October, 1992:261). Japanese plant breeders have developed two genetically engineered rice varieties that resist infection by the rice stripe virus. This virus, transmitted by insects, stunts the plants and reduces the amount of seed produced. By splicing a gene from the virus onto the rice chromosome the plant apparently produces its own vaccine against the pathogen. One of the scientists admits that 'the exact mechanism of protection is not really known . . .' but it works and will perform as well for family operations, the great majority, as for large commercial farms in Japan, Korea, China, Taiwan, and the former Soviet Union where the rice virus destroys a significant part of the annual harvest.

Similar genetic modification has also been conducted in wheat to address the problems of rust, a fungus that, like the rice stripe virus, reduces yield. The continued existence of rust on the wheat plant, however, has been criticized by some agricultural scientists as not getting rid of the problem 'once and for all.' We have already met this form of logic from earlier chapters. A fixed idea of definitive elimination of problems is not likely to succeed.

Grain diseases mutate as do all pathogens in continual adaptation for survival. The gene 'splicing' which increases yields for perhaps a generation is the kind of practical response that will guarantee food security that is needed now.

US midwestern corn and soybean producers are accustomed to making changes in order to use crop and livestock innovations. Some have only required technical modification. Setting in the spacing of corn rows a couple of generations ago to increase plants per acre, and adding more fertilizer, increased yields. Soon everyone did it and the 100 bushel corn clubs became old hat. Although the innovations in biotechnology are somewhat more complex, they also create opportunities. There will be a demand for crops, as well as for livestock, with specific characteristics. The 'special handling' that is often required is a strength of family agriculture. American farmers in Kirby, Ohio, provide an example. They raise one to one and a half million bushels of 'food quality' corn per year to be processed into corn chips and tortillas. Although yields are lower – the corn cannot be harvested with more than 21 per cent moisture – the premium paid by the company compensates for the extra work. This type of specialized requirement from industrial processors will become a source of established relations with farmers. Prices will be negotiated at levels depending upon requirements and may considerably exceed the local market price. The Kirby farmers also give an insight into the possibility of specializing as a community rather than as individuals. Although they are producing a specialized product, they can maintain diversified operations with livestock and, at times, part-time jobs off the farm. There will also be opportunities in the future to raise crops that go into completely new products. Oilseeds, canola or rapeseed, for example, will be used for differentiated oil products as well as biodegradable plastics. New market relations will be created through negotiations with farm producers.

As trade relations with the ex-Soviet countries have increased recently there has been interest in upgrading beef and dairy herds with imported stock. A representative of American cooperatives recently explained plans to establish an exchange of livestock and products. A large number of cull animals that will soon enter the market in the Slavic countries has attracted the interest of US fast food industries. Mixing the older leaner meat with the fatter American beef apparently makes a more profitable hamburger. Here too we can see the success of cooperation among family producers taking advantage of new opportunities to globalize their product. In collectivized countries such as Hungary (reviewed in Chapter 2) much of the livestock is produced on the private family plots. It is these producers who are organizing to put together a sizable number of animals to bargain for a better price in the market.

There is also a growing market in the developed countries for a secure supply of wholesome, natural foods produced organically without modern inputs. This demand is likely to increase and form a bond between urban consumers and farm producers. It is the type of link that not only provides an economic boost to family agriculture but increases understanding between rural and urban sectors and the life and concerns of each.

SPREADING THE WORD – THE EXTENSION EFFORT

With the imperative of food security and the complexity of biotechnology the need for agricultural extension is easily recognized. Transmission of new techniques, products and systems of production, through education and technical training, is more important than ever before. Even so, extension programs in both the developed and the developing countries are in decline, for a number of reasons:

1. The expanding volume of technical information has resulted in agents becoming generalists with some knowledge in many areas, to better serve their clientele. Industrial farms seeking specialized recommendations tend to look for solutions to agricultural problems directly from the universities or research centers. As they produce larger proportions of the national product, service to other (family) farms is considered less important.

2. In times of economic recession, a current global dilemma, public resources are limited. Voluntary educational programs typically lose priority and are cut back or even eliminated. Losses of rural population in the developed nations have reduced the traditional clientele of agricultural extension and urban residents don't sense the need for such programs.

3. Finally, the passing of colonialism and assistance programs in the developing nations has weakened fledgling national extension programs.

All this has come just when innovations on the market are increasingly complex. More farmers need advice on how to

reconcile time in off-farm employment with on-going agricultural operations and there is more need for farmers to join
together to compete with other structures of production to meet
the new specialized demands of the market. Brown *et al*
(1992:131) stress the importance of market information, a
traditional function of Extension Services:

> ... *if market information about production technology,*
> *market opportunities, and prices is not equally available to all*
> *market participants, society will produce less from its*
> *resource base than is technologically possible.*

The extension goal, in general terms, is to promote human health
and economic well-being through improved nutritional quality
of foods, more equitable access to foods, expanded economic
opportunities and better management of overall family resources
(Plucknett, 1990:35). Realization of this goal depends upon a
series of activities which directly involve primary food producers.
Looking back to the Middle Ages in Europe, there was little
occasion for innovative change in agriculture when, for example,
most physical labor was performed by slaves. The educated class,
who could have identified problems and elaborated solutions,
didn't get their hands dirty and the slaves were alienated from the
benefits of any improvements that they might devise. Thus
agriculture as a science lagged behind other areas of interest.
With the institutionalization of property rights efforts began to
be made to augment production. Over time scientific advances in
other areas were extended to agriculture with stimulating results.
When it appeared that agriculture could constitute a barrier to
national (principally urban and industrial) development, steps
were taken to integrate farmers into the fold. Busch and Lacy
(1983), discussing an initiative of the American Bankers'
Association in the beginning of this century, cite McConnell
(1953:31):

> *The [bankers'] program for agriculture included good roads,*
> *soil fertility, and education. Education meant cooperative*
> *demonstrations and county agents. This was the positive*
> *alternative to dissent and radicalism which the bankers held*
> *out to the sometimes misguided man of the soil.*

Thus began the Cooperative Extension Service in the United
States. In the beginning there was some confusion as govern-
mentally paid extension agents worked to organize Farm
Bureaus which subsequently united and began lobbying

Congress on behalf of the farm population (Busch and Lacy, 1983:25). As the conflicts of interest were resolved a principle was elaborated that has sometimes not been observed in the extension programs of other countries. Extension agents must be, and must be seen by their clientele as, completely autonomous. They must have the full confidence of the farmer so that any recommendations made are understood to be for the farmer's own benefit rather than for any personal, governmental or commercial gains. It is this client-centered orientation that has provided high levels of respect for extension programs and prestige for agents all over the world. In Latin American countries extension work began after World War 2 with 'a guy (an agronomist), a girl (a home economist) and a jeep.' Programs, typically subsidized by foreign aid, displayed picture albums of improved farming practices, water purification and related topics to increase agricultural production and improve life in rural areas. Soon the agents were addressed as *doutor* and *doutora* (doctor, masculine and feminine) and efforts were often made to please *them* as much as to improve personal situations.

Developing country extension agents often do not have the rural background and experience that has traditionally been considered essential in the developed nations. This is typical of countries where families who have the resources to send children to college are generally from urban areas and aspire to have their children become doctors, lawyers and teachers. When extension administrators determine that rural experience is important, agents are hired with less than university diplomas. Sixth grade graduates in West Africa are quite effective as extensionists and have no qualms about living in rural villages.

Agricultural scientists, too, who conduct the research that results in innovative farm practices, increasingly have limited experience in agriculture both in the developed and in the developing countries. As rural populations diminish there are fewer farm-raised young people for these activities. All the more reason for an effective Extension Service. Someone must represent to scientists the needs and interests of farm people, to direct public research where it is needed.

In some developing countries extensionists have become local representatives of state and federal governments with various duties that have detracted from their effectiveness in reaching extension goals. At one time they were charged with the supervision of farm loans. One agent reported that often when arriving at a farm the person responsible would be informed that 'the man from the bank is here.' With this reputation the chance of winning farmers' confidence is seriously jeopardized.

The friendly give and take between trusted extensionists and farmers is a characteristic of rural areas in many parts of the world. A farmer recently commented to his local agent, 'I'd sure like to have the money I've spent to prove you wrong!' He later admitted that he'd no doubt earned more from using extension advice than it had cost him.

In terms of food security and extension activities, we can summarize our argument as follows: family agriculture generates qualified, culturally sensitive extension agents who, through effective public programs, develop competent agricultural leaders, who, in the present situation of tremendous change in food production, processing and distribution are the best hope to represent the primary sector in planning for a secure supply of food, feed, fiber and agricultural raw materials for the future of all nations.

8

Land Consolidation, for Production or for Power?

Historically, determination of farm size has often been associated with family organization. Where extended families lived together they possessed the labor requirements for cultivating larger areas or tending larger herds. They interacted with their physical surroundings to develop effective methods of agriculture. For these reasons land consolidation is a topic of interest to us. The third question posed in our goals for the discussion of family agriculture asked how it can be strengthened to better the lives of farm people, and all of the population.

In the Amazon region, reviewed in Chapter 2, we learned of people planting as soon as the flood plain was dry enough, and working quickly to bring in the harvest before it flooded again. It would make little sense to plant more than could be cut before flooding. Natural constraints can, of course, be partially overcome by using technology. While an 'acre' was originally the amount of land that one person (and a horse) could plow in a day, now we plow one hundred! Slash and burn agriculture in Asia, Africa and Latin America is conducted by people cutting an area that they believe their families, or united family work groups, will be able to cultivate. Diversified yeoman family agriculture was developed in the western countries based on family needs first, and secondly the application of family resources in the production of goods for sale. Governmental farm policy also contributes to the determination of the size of structures a society establishes for its agricultural production.

COLONIALISM AND THE RISE OF NATIONALISM

Colonization by the European countries in the developing areas of the world changed the traditional structures of agricultural production. The establishment of estates or plantations produced large areas of monoculture for export to the central powers. Land for plantations was often allotted, by the governments of the colonizing powers, to notable citizens of their countries who, they believed, could 'domesticate' the areas. The objective was to increase the possessions of Europe as well as to produce useful goods for export to the home countries. These citizens received enormous tracts of land with nearly unlimited control to do as they liked, as long as they produced results for their patrons. In the Americas the plantation system that ensued was established for the production of specialized crops: tobacco, sugar cane, cotton, and later rubber, coffee, cocoa and palm oil, that could not be grown in the temperate climates of the developed countries. For labor requirements the local population was seen as an initial resource. When the native Americans did not demonstrate the will to work as the colonizers demanded, waves of African slaves were imported along with European, indentured servants to assure an adequate supply. Adjustments were made to the conditions of the rustic territory, and production increased.

The plantation system served as an agricultural structure that supplied the needs of the developed countries from as early as 1550. In the nineteenth century this system was spread to Asia and competition was established with the Americas to supply Europe. In the twentieth century Africa also was incorporated with, as an example, the establishment of Harbel, Firestone Rubber Plantations in 1926, in Liberia.

It has been claimed that the plantation system has had positive effects on developing areas, as explained by Mary Tiffen and Michael Mortimore (1990:132):

> *It seems likely that plantations play their most positive role in the early stages of economic growth, when they provide an important source of foreign exchange and a taxable capacity, which can be used to build up general infrastructure and services. At this stage there is likely to be under-utilized land and labour in the traditional agricultural sector.*

But those times have passed. Local populations benefited from the schools and examples of economic growth that were implanted in their countries. They began to make demands on the colonial estates. As populations increased there was more need for land to produce food crops. The plantation was perceived as a foreign structure imposed on local people. With the rise of nationalism around the world it became unacceptable, in many areas, to have foreign owned lands producing for export. The local advantages of the plantation economy went principally to a small minority of the local population who, using their wealth and power, were vigorous in attempting to convince the rest of the population that the system was not only beneficial but a protection from the dangers of local corruption and chaos.

Today populations in the developing countries are not so easily hoodwinked. It would be very difficult to consolidate tracts of land to produce for export, even if the initial force came from the local government. An attempt in the Cerrado region of Brazil during the 1980s to establish large areas of soybean production for export to Japan, in spite of financial support from the Japanese, produced little success. Rumors quickly circulated that thousands of Japanese migrants would come in to take over the region and exploit resources for the good of Japan. While this was never the intent of the program – the producers were all Brazilian – it created enough controversy that the program became a political liability. Those who received the land paid their debts in soybeans (assisted by high prices) and proceeded to raise cattle or whatever suited their own family interests – though some did continue to raise soybeans.

The possibility of contracting with private producers for the raising of export crops is the solution that past plantation owners have found to perpetuate the international flow of produce in spite of changing socioeconomic conditions. With this change from consolidated commercial structures to smaller, generally family structures we have an opportunity to compare the two forms of production.

It is undeniable that the periods of 'development' in the colonized countries, during which plantations were established, produced laudable results in terms of reductions in infant mortality, increased life expectancy and capital formation among others. But those benefits are forgotten today as reports come out of losses of renewable resources and the polarization of rich and poor, both within and among countries, attributed to the plantation economy. The result has been ecological crises, population growth beyond resource increases, and enormous debt. It is not surprising then that local interests turn inward to self sufficiency and resource protection. Tiffen and Mortimore, citing research on the topic (1990:21–2), report that in the developing country situations it is felt:

- that export agriculture has for too long received priority in investment, subsidies, research and other policies that now should be switched to food production;
- export agriculture weakens the position of the poor by competing with domestic crops for scarce resources and by encouraging land concentration; and
- export production develops weak linkages with the rest of the economy, restraining general growth.

   Even such linkages as may develop (employment, fiscal, consumption, technological, social and infrastructural links) often act to favour the export enclaves at the cost of the subsistence, indigenous sector of the economy.

It is true that some East African nations recently began inviting foreign ex-land holders, expelled during the processes of independence, to return and reinforce the production sector. These efforts are now being called into question, however, as courts must furnish clear positions of land rights. On the other hand, there is no lack of interest on the part of the traditional colonizing countries in returning. Tiffen and Mortimore write (1990:15) that the European Union countries as well as the British Overseas Development Administration have funded and would like to participate in more plantation developments.

In some countries the passage of stringent labor legislation to satisfy the majority, working segment of the population has put the plantation at a disadvantage. Being highly specialized, plantations have limited possibilities to switch crops or methods of production in response to changing conditions. With the objective of producing a specific crop for a specific market, they frequently act in disaccord with the local situation. They also face the possibility of unfriendly local governments that might increase their taxes, limit the amount of profit that may be converted into dollars or other foreign currency to leave the country, and even nationalize the lands, taking them away from the foreigners for redistribution to local citizens.

Tiffen and Mortimore explain that the advantages of the plantation structure are not in economies of scale, the cultivation of large areas to effectively exploit resources, but in managerial efficiency, combined with adequate levels of capital. This makes it possible to do more experimentation, introduce modern technology, respond to market conditions, maintain installations adequately and get full use from resources. This is not to say, however, that all plantations are efficiently managed. Since, as we have seen, they are more likely to be justified in situations where there is an agricultural frontier, there is little competition. Today the agricultural frontier is gone all over the world and, as a result, there is more interest in existing agricultural lands and their patterns of use. If the plantation uses extensive areas of land on a low input basis, it has no advantage and is actually inferior to a family producer who has limited capital, but is more likely to be keenly interested in protecting the land and other resources.

It is these emerging problems that have resulted in plantation organizations negotiating with individual, local farmers. The use of contracts to assure production also leaves the exporters with less capital invested in land, buildings and equipment, and less managerial responsibility so that more emphasis can be given to processing and marketing. At the same time the exporters are less vulnerable to criticism that they are consuming local resources in favor of external interests.

While this example comes from the specific characteristics of international relations, it also demonstrates the breakdown of a large, well financed agricultural structure and its replacement by smaller, more flexible, privately owned farms.

Some of the most progressive of the developing nations have maintained parts of their plantation structures, but adapted them to serve local needs. In Brazil, for example, the cane, rather than being exported for sugar, is being processed locally into ethanol to partially substitute gasoline, which for the most part is imported. In such a case, Tiffen and Mortimore point out, there may be significant possibility for local growth and the development of an industrial sector. In the case of food products this is especially valid. Still, in comparison with private, family based production, the authors conclude (1990:133):

> ... it is likely that there is greater demand from the small-holder sector for consumer goods and services which can be produced locally, and that the demands from the estate sector may, at any rate initially, be more orientated towards foreign sources.

Families are more likely to spend their incomes in local communities, strengthening them and promoting development. While they do not practise such specialized forms of production (that would be risky), this is not an indication that they are less efficient, since resources are combined in different ways to produce diverse products that are probably more in keeping with the needs of local people. Another aspect, often discussed in my classes with veterinary students, is the alternative options for employment between a few large farms, plantations or locally owned, and many smaller, family owned operations. Which will need more technical assistance? Which can support more veterinarians? Just as families spend their incomes on local goods, they also support local services that increase interaction and fortify the technical expertise available in the community.

THE COMMUNITY: A SUM OF ITS PARTS

Large farms in developed countries have characteristics somewhat similar to the modern plantation. They specialize their production and rely heavily on hired labor. Cornelia and Jan Flora studied the effects of farm size on rural communities in 1200 counties of the North American plains and western states (1988). They wanted to identify the effects of large (2000+ acres) and medium sized (500–999 acres) farms on their local communities. Their findings demonstrate that the communities that had larger quantities of sales generated from retail establishments were more likely to be located in counties with higher proportions of medium sized farming operations. They called this a *measure of community vitality*. It was attributed to a series of factors (1988:117):

- Medium sized farms have a higher capital-to-labor ratio, utilizing more labor per unit of product produced and thus keeping more people in the rural area.
- These farms see it as an advantage to make use of labor rather than capital (perhaps borrowed) and resist, for example, excessive mechanization.
- They tend to buy locally, both in terms of agricultural inputs and household needs. They also sell their products locally, increasing the flow of capital in the community.
- Medium sized farms are more likely to be operated by their owners rather than have absentee owners with little or no interest in the local community.
- Local interest and participation generate new economic activity to serve local needs.

Large farms, on the other hand, were similar to plantations in purchasing and selling in distant, regional or national markets. They were less integrated in the local communities and depended, for labor needs, on temporary, migrant workers. Thus, although the land of the community was occupied, the labor and capital bypassed local possibilities for the development of adequate community services. Large farms thus mean smaller communities, and less available labor for industry that might be interested in locating in the rural area.

In examining land consolidation these researchers' attention was called to the significance of levels of farm mechanization. Farmers who have invested heavily in machinery will, it was thought, have been highly motivated by possibilities of increasing their areas of land. But investments in machinery were found to be better explained by the ease of its acquisition, because of negative interest rates and tax policies. Only subsequently was there concern to increase the area cultivated. This type of tax policy works to the disadvantage of typical, middle sized farm families that don't have such large investments in machinery. Tax policies that allow large writeoffs of nonfarm income also work against the family who doesn't have much of this kind of income. They favor, in this case, the part-time farmers (Flora, C B and J L, 1988:123).

These findings raise doubts as to the validity of criticisms that family agriculture is less efficient than other production structures and requires modernizing. Other research has found similar instances of policy decisions which have stressed growth without regard for the consequences. Lawrence Busch *et al* (1991:54) reported that:

> *A recent study, examining this problem in the United States from 1915 to 1973, concluded that public agricultural research significantly increased farm size during that period independently of other contributing factors (eg, debt, taxes, unemployment).*

Recent research in some developing countries has resulted in efforts to produce methods and technology that require less capital rather than less labor. Emphasis is thus put on increases in product/acre or per unit of investment rather than per hour or per 'man unit.' The evaluation of family operations, from this perspective, is completely different, and generally positive.

Results of the investigation conducted by the Floras (1988) call to our attention the importance of examining family agriculture in relation to the rural community and reinforce the relevance of improving living conditions for all rural people, to be addressed in the next chapter.

LAND FRAGMENTATION, RISKS AND ECONOMICS

In the older societies of the world agricultural lands have generally been fragmented so that one family has several different plots around the community. In Greece there are plots so small that they have space for little more than one ancient olive tree. Agricultural scientists have typically criticized this fragmentation as being inefficient since it makes mechanization difficult, wastes land for fences and roads and requires extra time to travel from one plot to another. They have encouraged programs to redistribute family lands, consolidating the total acreage into one parcel that is approximately the size of the various plots. Still, the system persists in many different countries, from Eastern Europe to Japan and Africa.

Now research has been ordered by the World Bank (Benoit Blarel et al, 1992), to achieve a better understanding of this traditional system. A group of technicians surveyed the findings of previous studies and collected farm information to identify producers' attitudes and opinions. They found that the various plots families cultivate are generally in areas that are somewhat different in their climatic characteristics. This often provides protection against the local risks to production. Where the problem is water, for example, farmers may have drier plots, safe from flooding, and lower, more humid areas in case of drought. In other regions the problems may be fire, pests, disease, stealing or any of the many risks that farmers face. In these cases too, the use of various plots in different areas diffuses the risk. Some soil types may be better for certain crops and it is to each family's advantage to have at least a little land of each type, to diversify production and assure income. There is also the possibility, by having part of the family's land in areas that can be planted and harvested slightly later, of stretching family labor to cultivate more land. This information prompted the research team to admit (Blarel et al, 1992:234):

> Although there are negative aspects to farm fragmentation, there are also reasons why it may be beneficial to farmers.... In fact some degree of farm fragmentation may be desirable and might best be viewed as a rational response by farmers to the economic and institutional environment in which they live.

And what of the consolidation programs that were to modernize the local agriculture? 'The conclusion is that consolidation programs are unlikely to lead to significant increases in land productivity and may actually make farmers worse off.'

To put the question of farm size in perspective we need to remember that it is a problem associated with private property. In West African villages where elders distribute regional agricultural lands periodically, it is much simpler. There is also the system of the Tamberma in northern Togo, who cultivate areas of similar sizes in the belief that increasing the area planted would be asking for more than one's share of the region's harvest. Then, when a family's granary empties during the dry season, they just move in with a brother or other relative who still has food!

Explanations of farm size vary among nations. There are socioeconomic factors, in addition to the basic determinants listed at the beginning of our discussion, that are fairly universal. Others are particular to each country. Land as a value reserve, for example, is universal. Individuals with extra capital, looking for investment opportunities, are found in most countries and not only where there is already developed agriculture. In some nations the possibilities for retaining wealth are limited. There, land ownership is a source of security. Many countries offer tax shelter advantages, making it possible to develop farm land for a fraction of normal costs. There have been cases in which large farming operations have developed frontier lands using 'purchased' livestock and machinery from their own farms and abandoned or given them away when the government subsidies expired. Some industries, even multinationals, have purchased frontier lands in developing countries as a promotional activity to demonstrate that they are interested in the nation's future. Others decide that by controlling the production of their own raw materials they can produce what they need at less cost.

These are all explanations for the existence of large farms. We might also ask, then, why haven't the agricultural sectors of all countries evolved to large farms? Some have had farming for thousands of years and have not significantly changed in size.

The same industries that decide to 'vertically integrate' and produce their own raw materials may decide, at a later time, that it would be cheaper to contract this production and not have the responsibility for labor, the investment in land, or the necessity of accepting the product that they produce, even if the quality is inferior (Mottura and Mingione, 1991). They may come to the conclusion that the same amount of capital, invested in the industrial sector, would generate more returns. They, like the plantation producers, thus seek alternative methods of acquiring raw materials.

The agricultural sectors of most of the older societies have come to represent stability in their nations. In Kenya after World War 2, British planners, working with local technicians, sought methods to stimulate national development. Anne Thurston (1987:70) explains that after many frustrated attempts, '... agriculture was increasingly seen, at all levels of government, as the means of achieving the most speedy recovery and a return to order and control.'

A plan was elaborated to use policy incentives for shifting emphasis from large scale European farming together with subsistence African agriculture, to commercial African agriculture. Family operated units were established of a size that was economically feasible to produce for national needs, still maintaining some of the traditional subsistence orientation. These farms remain the dominant type of production structure in the nation today.

Large farms, on the other hand, can present special problems. Lawrence Busch, sociologist at the US Michigan State University, tells of a work visit he took to Australia where local university personnel showed him some of the characteristics of their large farms: considerable distances from household to household, with the costs of communication and electric lines and, in some places, piped water due to the presence of salt in the local water. Costs just to provide basic services could become unsupportable. The 'smallholder' agriculture of other areas, where many families are in more immediate proximity and share the expenses of community services, is comparatively a lot cheaper.

Another factor of increasing importance, related to land consolidation, is employment. As world population continues to grow, the labor saving orientation that stresses increases in production per hour or per person is less and less appropriate. We need to produce jobs just as we need goods and services. Large farms aren't useful from this point of view. Although the plantations of developing countries were based on the intensive use of labor, they are now disappearing, or modernizing to increase labor efficiency. Smaller family farms, on the other hand and as we have seen, have demonstrated the capacity to absorb great quantities of labor in times of necessity. Hartmut Schneider (1984:101) cites a study conducted in 40 countries which concluded that farm output and employment per area of land 'tend to be higher in countries with smaller average farm size and more equal distribution of land.'

A final point, still unconsidered in our discussion, is the politics of farm size. We learned in earlier chapters that democracy grows from the bottom up, and works best among individuals and communities that are fairly equal in their socioeconomic standings. This has implications in communities that have a mixture of farm sizes. From a US data base, Linda Lobao (1990:72) states that:

> The effects of farm structure and change extend beyond the farm gate, reverberating across localities to the extent they depend upon this economic activity ... generalizing from a political economy perspective, large, industrialized farms should lower socioeconomic conditions and lead to greater inequality in localities. Family farms should have differential effects on inequality: those sustaining independent producers through farming alone should benefit localities; those that do not may have adverse impacts ... however, these relationships are not immutable but depend on wider social and historical conditions, including popular action.

CONCLUSIONS

What can we conclude from our discussion of farm size? We've discovered that size is determined by the production structures used by each society, whether it applies family, plantation, industrial, collective or a mixture of organizations. The geographical and climatic conditions are important. Technology has made changes, and government policy has a lot to do with it.

In many cases it matters little what we decide. Natural (climatic and geographical) factors may overwhelm technology. In other cases politics may be dominated by one agricultural structure or another, to the disadvantage of the others. But in the interest of a democratic goal (remembering consumer interests too) we can suggest a position that would be most beneficial to the largest proportion of the population.

Perhaps the most appropriate organization for agriculture, as a general, cross cultural suggestion, would be a mixture of large farms that specialize in products that lend themselves to industrial processes, for example, eggs and broilers, and are located in areas where they can make efficient use of local resources; and a larger, dominant part of production handled by families, integrated in as much cooperation as possible, and working in manners that strengthen rural communities and complement, rather than compete with, the industrial structures. In this way we could hope to build a sector that has more chance of working together to achieve fair government regulations and opportunity for all.

9

Agricultural Policy:
Shell and Pea Games

Agricultural policy has historically been an elusive entity; now stimulating production, now abandoning farmers for other priorities, all over the world. There has rarely been focused national planning to direct organized policy over extended periods of time, as has been common in the industrial sectors. To make it even more difficult, the nature of agricultural production requires time for changes in farming operations. This is especially true of family farms where capital is individual and limited. Still, agricultural policy has the potential to be the most important factor in strengthening family agriculture in all areas of the globe.

POLICY FOR WHAT?

Maybe we should do away with policy and live by simple supply and demand for our products. We could produce whatever we wanted and increase our independence ... as long as the machinery and equipment dealers and the seed corn people and the grain elevators and the banks didn't do the same thing, and other countries didn't flood our national market with cheap products – in that case we might end up on the wrong end of the bargain. Perhaps we should feel better – or worse – that the primary objective of agricultural policy has been questioned for more than 600 years! D G Barnes wrote in *A History of the English Corn Laws* (1930:3), that:

> The policy as a whole shows no consistent purpose during the three hundred years before 1660. At one time exportation would be forbidden in order to ensure cheapness and plenty for the consumer; and at another time it would be permitted in order to help the producer dispose of a surplus.

Still, for all of the policies and programs and subsidies and supports, the agricultural sector is not protected from destructive change that destroys lifetimes of hard work and savings. The farmer pays industrially determined prices for the inputs purchased and receives industrially determined prices for what is produced. Why do we really need policy? The answer is that policy is a political instrument to stimulate beneficial activities and direct or redirect the distribution of resources in line with the values and beliefs of the society. Marty Strange (1988:249) recommends that US farm policy be elaborated in the form of a 'modern economic and environmental land ethic':

- A farmer should be able to pay for farmland by farming it well.
- A farmer should have to farm it well.
- A farmer should have to pay for land by farming it, and by no other means.
- There should be no other motive for owning land other than to make a living by farming it well.

So what do values, beliefs and land ethics all have to do with family agriculture? The farm, as we have seen above, symbolically represents what is pure and honest in most societies. The family typifies ideals of working together, diligence rewarded in the development of character, and conservation as an indication of faith in the future. Of course today in the developed countries this is commonly seen as myth. In practice we're just trying to keep body and soul together and meet the next demand for payment. But society continues to attribute an archaic, sentimental significance to what we do. Why is this? A simple answer is that the myths of farm life persist because they express basic values in developed societies too, and give them meaning. The farm family, for example, furnishes an answer to 'what is a family?' Differentiated, complementary roles, interdependence, vulnerability, protection and security, as discussed in Chapter 6, are all exemplified in rural family life. The modern, industrial 'throwaway' society, on the other hand, is imbued with aspects that are not so admired. When there is a need to work a distance from home, or in several different places, there arise conflicts of hours that disrupt family togetherness. Life becomes disarticulated as members are not aware of many of each other's activities. A woman recently told me that she uses stories of her childhood on the farm to entertain her grandchildren. Now every time the kids come over they ask her to repeat the stories. Farm life is associated with 'lives and deaths' and the vital issues that are interesting to children and to all of us. These issues, even as they stimulate communication across the generations, are value laden. They exemplify how we react under given conditions and give us content to fall back on.

Today, in the developed countries, very few people live on farms and stories of rural life are harder to come by. But the aura is still remembered, and the society both uses policy to express these sentiments and uses the sentiments to formulate policy content.

POLICY OBJECTIVES

In determining objectives it is important to remember the relations that exist among the sectors in each society as well as among nations. At the same time that there is interest in maintaining farm income at reasonable levels there is also interest in food costs in the urban sectors, foreign trade relations and service to humanitarian causes.

Governments establish agricultural policy objectives in relation to specific targets. These measures, too, are able to achieve success only through their interrelations as they reinforce each other in addressing various aspects of problems. They typically include the following:

- Price stability of agricultural products
- Low farm income
- Food security
- Food safety
- Future agricultural production capacity.
  (*Taken from Knutson et al, 1983:12*)

We will now examine each of these objectives to identify its relevance for family agriculture.

Price stability

Governments are particularly sensitive to the need for stability of agricultural prices because it affects both producers and consumers. At the same time that low prices reduce farm income, high prices reduce consumption, provoke dissatisfaction among salaried workers and hurt those in poverty. Still, the possibility of regulating prices effectively is a luxury of the wealthier countries only. It is a burden to those in the process of development, and an impossibility in the poorest nations.

The means by which different segments of the population are protected against changes in prices vary. Support prices, when established by the government before planting time, help producers make decisions and control risks. In times of economic recession, however, it is difficult in the developing countries to secure commitments from the bureaucracy in time to begin soil preparation and planting. Recession also affects the availability of credit and has direct effects on production levels.

With the establishment of support or 'government' prices, it may be necessary for the government to make purchases of the product to reduce the quantity that is being traded in the market and sustain the price levels guaranteed. In this case the product must be stored for release on the market in the case that prices become inflated (this is an option of the wealthier countries that have storage facilities), or it may be disposed of through programs to assist those in poverty, either at home or abroad. Milk distribution and school lunch programs are examples. These programs can, however, have a somewhat depressing effect on the national market for the products involved. In the case that the product is shipped to a foreign market there may be an export tax levied on producers to offset costs and the possible loss of selling at lower prices. Such exports will most likely affect the market at destination and may create the problems discussed earlier with P L 480 wheat from the United States. Of course, if the support prices of a nation exceed world market prices there will be a need for import barriers to avoid flooding the national market with cheaper, foreign product.

On the consumers' side the government may establish price controls to avoid inflationary effects. The producer, whose principal interest is not price but income, may, for example, thus be forced to sell off brood sows in response to controls on pork, or severely cull beef or dairy herds when they are the focus of controls. The government may also restrict exports, forcing national producers into the domestic market, or more simply, in the case of products that are traditionally imported, ease the restrictions to facilitate entry.

The types of measures here discussed to protect producers have a disadvantage to family agriculture and to agricultural production in general: they tend to increase production. Price supports, for example, create conditions of stability that are a powerful stimulus to families interested in augmenting income,

and practically guarantee profits to larger than family operations. The possibility of more income or profit, in turn, increases
demand for farm land causing land prices to appreciate and
restricts the entry of young families into agriculture or the
expansion of existing farms.

Low farm income

Farm income is a source of perpetual concern by governments
since the responsibility for the national food and agricultural
produce is critical and those involved in its production need to
receive incomes reasonably equitable to other workers in society.
Even the poorest countries, while lacking resources, provide
incentives to agriculturalists to encourage production in accord
with national needs. Income supplements have been attempted in
some countries under various types of agricultural policy. They
have been used to maintain a working population on marginal
lands, avoid rural-urban migration, equalize rural and urban
income and for other purposes.

The 1992 reforms in the Common Agricultural Policy (CAP)
of the European Union nations shifted emphasis away from
increased production and productivity to secure income for
producers. A system of direct payments to farmers is based on
animals produced and hectares planted. Payment levels are
determined by average productivity in each homogeneous
region. Individual motivation to maximize production using high
levels of inputs (which would be undesirable) is thus reduced
since, although more product would be produced for the market,
payments will not be affected. These payments will increase up
through 1995 to reach a full 30 per cent of crop value. At the
same time, product support prices will be reduced (by about the
same amount) over the three years. Pretty and Howes (1993:46)
provide details:

> EC-wide limits on crop and livestock production were set and,
> should farmers exceed these limits, they will be penalised in
> future years by having to comply with more stringent controls
> on production. Market prices have been reduced, and finan-
> cial support and incentives linked to specific farming practices
> introduced. The incentives would appear, therefore, to be
> present for farmers to comply with new practices, and so
> reduce food production. Sustainable technologies and
> practices represent an element of these compliances.

The controls already introduced for farmers to participate in the program include setting aside 15 per cent of cropland on a rotational basis. Although this measure provides immediate results in terms of reducing production, it also reduces needs for rural labor. Pretty and Howes cite an estimate that one agricultural worker will lose his or her job for every 130 hectares of land set aside. In England alone this would have resulted in the loss of 4500 jobs in 1993. Forcing more farm people off the land strains rural community services and threatens community vitality.

Payments associated with reduced planting and reduced production also run into problems in terms of support from the urban population. There is a negative reaction to policy which advocates transfers of cash in exchange for less work and less product. There is also a negative reaction on the part of farmers who feel that land set aside is a lost opportunity and excessive control.

The question of size of the farm population has cultural dimensions as well as being influenced by aspects of topography, mechanization, and other factors. Scandinavian farms still have only reached around twenty hectares, are actively diversified and increasingly complemented by off-farm income. But mechanization has been an infatuation in the New World – both in North and South America – and the labor saving advantages have often been adopted in Europe as well. With mechanization has come increased isolation, less working together and, with time, more individualism. This has meant less interest and ability to join together for mutual defense. The US farm population was reduced during New Deal times of the 1930s and this trend has continued. Ironically, family agriculture in the Midwest was actually strengthened for a time, as capital from New Deal benefit payments was used for mechanization (McClellan,1991).

Today American farm income is directly correlated with export demand for agricultural products. This is generally acceptable as long as export demand is sufficient to provide adequate levels of income. But it should be seen as somewhat precarious that the income of an important sector of society is determined by external factors. Knutson and his colleagues (1983:188) admitted that 'Reduced exports over an extended time period have the potential for creating resource adjustment problems as severe as those experienced in the 1960s.'

Food security

Agricultural policy directed to food security involves the problems that we have examined in Chapter 7. For most countries this means avoiding food scarcity. As noted, a large part of the populations of developing countries spend half of their incomes on food and don't consume meat more than once a week. But in all countries food demand, reflected in prices, is a topic of economic importance as families make weekly purchases and are quite aware of price increases. In Brazil, for example, the inflationary situation is quantified, for most people, in the price of bread. Daily changes are noticed and provoke reactions. For this reason the politicians are particularly attentive to adequate food policy. However the Brazilian Government itself, the major purchaser of milk to be redistributed through a food stamp type program, has been accused of fixing low prices to reduce its cash output.

Poorer countries and the poorer segments of even the wealthier countries receive special attention from governments as well as from numerous national and international organizations who provide 'baskets' of food where there is chronic need, in times of crisis, and during holiday seasons.

Food producers, although in general protected by agricultural policy, no longer receive the major part of the food budget in the developed countries and their part is decreasing in the developing countries too. 'Value added' processes offer convenience and reduce time for food preparation. Knutson et al (1983) suggest that the farmers' shrinking share of food prices is not a cause for concern, however, since even when farm products are sold directly from producers to consumers the benefits go principally to the consumers, who must then handle the preparation. The farmers' share does not significantly increase.

As far as security from world famine is concerned, as was widely predicted in the 1970s, that 'future disaster' is already in progress. War, starvation and disease in the poorer countries and the acceptance of these events as either normal or beyond control in the developed countries has become common. The populations of these countries will have difficulty to avoid being reduced to levels that can be supported by their reduced resource bases. The fabrication of industrial foods together with productivity increases of recent decades in the developed countries suggest that future food problems, for them, will not likely be those of traditional food security but because of inadequate research and diffusion of innovations. Hopefully with time these benefits can also be integrated into survival programs in the poorer countries to establish more promising life conditions.

Food safety

Federal inspection of agricultural products is undoubtedly a safeguard to human and animal health in all countries. Control of the spread not only of pathogens like hoof and mouth disease but also Mediterranean fruit flies and DDT residues, is necessary. Many producers, however, see regulations on growth stimulants and pesticides as unnecessary interference from the government. The regulations established by most governments are stringent and rigidly controlled. On the other hand, the same requirements are demanded of imports. The US, for example, will not import meat from nations where certain communicable diseases of the animals are present. In the approved countries both slaughterhouses and carcasses are inspected for parasites and chemical residues. Fruits and vegetables are also examined for residues. So at the same time that there is less than independence in crop and livestock production, there is also protection from imported toxins, pests and diseases.

Economic conditions in developing countries, it must be admitted, are influential in national disease control programs. Paraguay, for example, has government veterinarians who conduct intensive vaccination programs against hoof and mouth disease at no cost to livestock owners. Still, the disease remains out of control and the possibility of eradication nonexistent. A study conducted some years ago by an extension agent (Medina, 1977) interviewed producers in an attempt to discover methods of improving the effectiveness of the program. He learned that producers were well aware of the disease and how it develops. They also knew the effects of the vaccination. Thus when the government veterinarians arrived at a neighbor's property, working their way down the road, they hid weaker animals that they feared would not withstand the vaccine in woodlots or remote areas of the farm 'until the danger passed.' The rest of the herd was vaccinated regularly and general cooperation was at high levels. But the disease persists.

The control of farm pests with agricultural chemicals will, no doubt, some day be greatly reduced or a thing of the past. Natural controls and genetic modification generate less pollution and undesirable after effects. Prohibition of chemical use would necessitate more labor intensive measures, as are already being used in some situations. Termites and ant colonies that consume enormous pasture resources in the tropics and sub tropics, for example, can be destroyed by digging out the 'queen' and removing her to stop reproduction processes. Insect traps in orchards and the oiling of cattle reduce parasites if performed regularly. It is probable that the obligation to use more labor would favor family agriculture over industrial operations as quality and diligence in the task are required.

Future agricultural capacity

Government policy is especially important in insuring a nation's future production. There are alternative positions that can be taken: the government can minimize its influence using the strategy that it should not interfere in individual farmer decisions. Free market forces would thus be relied upon to reward those who conserve soil and water resources for future production, with better land prices. Farmers would also be free to respond to short-run market demand by temporarily intensively exploiting resources to raise needed cash. Or the government can become an active influence, recognizing that resource conservation is not only an individual interest and responsibility but has critical social consequences and the society must be willing to enter into partnership with the farmer to develop other feasible economic choices.

Soil and water conservation are not economically remunerative to agriculturalists in the short run. As decisions are made in response to market forces we must realize that the market is not a mechanism with social conscience or concern for the future. In a class exercise, my veterinary students are asked where they would buy corn to feed cattle: one producer rotates his crops, controls soil erosion and doesn't pollute downstream water resources. Of course his corn will cost more, about sixty cents a bushel, but he can supply it on an indefinite basis. A second farmer raises continuous corn, applying whatever inputs are necessary to realize short-term profits. The future does not concern him at this time. Generally there is a clamor in the classroom to buy this second producer's corn. For the farm owner himself to shoulder the costs of resource stewardship in the hope of some day selling his land at higher prices is impractical. Production decisions and income needs are not focused in such long range terms. Also the question to conserve or not to conserve is not an either/or decision. Most of us sense the need to conserve at the same time that we cut corners to get by as best we can. But the category of producers who can be counted on to have other motives, to protect resources and think beyond short-term gains, has already been well identified in this discussion. This category contrasts sharply with industrial farm managers who are required to produce annual profits in spite of labor, market or weather conditions and will consider conservation only after other priorities have been met.

Another problem of depending on free market forces is that in countries with unstable economies, as we saw in the last chapter, land serves as a value reserve, a place to speculate for future profits and invest capital that will inflate with inflation and maintain value. In this case questions of production are secondary, much less questions of conservation.

The decision as to whether a government will adopt passive, free market, or dynamic, partnership policies of resource conservation is strongly influenced by those who formulate policy. Unfortunately the worried soil scientists mentioned in Chapter 5 are less likely to participate in policy making than are economists who seek to generate revenues to apply to the nation's economic dilemmas. There are, however, some regions where environmental damage can no longer be ignored. The Sahara Desert, stretching across West Africa, is descending toward the Atlantic coast at a rate of one kilometer per year. Desert winds have incredible force. The desert 'grows' by storms blowing a few feet of sand on top of 'low bush' and the vegetation is gone. Recuperating such areas will call for intensive agricultural policies (as the Israelis have demonstrated in similar areas) and other resources so that, for example, people in the affected areas have no need to cut trees out of windbreaks for firewood to supply household energy.

For resource policy to be effective it is essential that all residents of target areas work together. A water and forestry resource manager from India recently explained the linkages of the upstream and the downstream populations. Those in the downstream areas must do more than complain of siltation and pollution. They must, he declared, join the upstream people in paying for changes in their cropping systems. The downstream effects are not so easily negotiated in areas that have a mixture of industry and agriculture.

To be sure, the most valuable resource to be protected for

future agricultural production is the producer. The knowledge and techniques of producing, sometimes in areas of marginal physical characteristics and climatic irregularities, have been passed down through generations together with the dedication and will to make them work.

As an example of how agricultural policy is being dynamically applied to address circumstances of changing need, we will refer again to the writings of two authors from Sweden and Norway that were reviewed in Chapter 2.

FARM POLICIES IN SWEDEN AND NORWAY

The Swedish experience

Sweden was reluctant to join the European Community following World War 2, fearing a loss of autonomy and control over its own national priorities. One of these priorities was its agricultural structure. This fear was reinforced at the 1986 Tokyo economic summit which acknowledged that the 'global surplus of grain production was no longer tenable' (Vail, 1991:256). With its own surplus of roughly a million tons of grain per year, Sweden saw no advantage in getting into a larger predicament. Since it is a relatively small country, national leaders determined to resolve their problems within their own territory. Political leaders and interest groups, while starting from divergent points of view, have worked out elaborate compromise solutions. Vail (1991:259) explains:

> The shift in Swedish public and scientific discourse about agriculture is truly remarkable. Every industrial nation has its agrarian movements promoting family farms, chemical-free methods, animal health, landscape, and other non-production values. Sweden appears to be unique, however, in the extent to which these values and ethical convictions have been mainstreamed into the consciousness of citizens and the platforms of politicians – shifting from the periphery of policy debates to the center. Swedish agricultural policy is in the process of being reformed along lines that would be seen as economic lunacy by the major players in American farm policy debates.

Vail suggests that the results have been positive in reducing agricultural surpluses (he cites the 'butter mountain' and the 'wine lake'), regional farm income disparities, and rural unemployment and underemployment. These results have come from a combination of agricultural, rural development and labor market policies. They began by recognizing that

> ... *price supports have the static effects of stimulating supply, depressing demand and inflating land values. On the other hand, price-based income supports slow the withdrawal of land, labor and capital from agriculture. These policies, combined with a continuous flow of yield-increasing technological innovations, are a recipe for structural – as opposed to merely cyclical or episodic – overproduction.*

Commodity price control was established by the government along with the use of selective production quotas. A campaign of moral persuasion to reduce production has also had positive effects. Success of the program has been attributed to the participation of the 'Cereal Grains Group' of farmers and the National Farmers' Federation. Their political force, in spite of falling numbers (farm labor force was 3.7 per cent of the population in 1980), has been reinforced by the food processing interests. These industries were created by farmer cooperatives and now employ more people than agriculture itself. Vail (1991:265) also cites some researchers who have pointed out the existence of a 'citizens' reservoir of good will':

> *Sweden's relatively recent urbanization, a widespread love of nature, and city dwellers' personal contact with farmers during their country vacations all contribute to the positive cultural valuation of agriculture.*

Agricultural policies are still evolving to formulate more adequate solutions. The agricultural groups themselves proposed the idling of crop land through short-term fallowing and tree planting. Environmental interests were joined by agricultural scientists, however, in responding that it makes little sense to idle some land while continuing to apply chemical intensive methods on the rest. Taxes on chemical fertilizer have been increased since a series of fertilizer-related fish kills in the Baltic Sea in 1988. Other suggestions include the production of protein fodders (field peas) to substitute imported protein, grass-fed beef and research on bio-mass energy crops.

Farm policies in Norway

The Norwegians also rejected European Community membership in 1972. It was stated that 'full community membership would have struck a severe blow to Norwegian agriculture because of its structure and climatic conditions' (Almaas, 1991:278). Agriculture has traditionally been combined with fishing and forestry and, more recently, with industrial occupations as families work to improve their standard of living. The decision regarding Community membership was associated with an extensive national discussion stimulated by the studies of three professors of agricultural sciences. Almaas (1991:280) says that their research showed that, essentially:

- with rising productivity from high levels of inputs of biological and mechanical technology, farmers would have to leave agriculture because food consumption is fairly constant and the population stable;
- agricultural and fishing policies which favor larger than family operations over smaller ones would hinder individuals who adapt their work schedules to participate in both areas, reducing the income and standard of living of all who are forced to specialize in one or the other of the two occupations;
- transfers of capital to agriculture have not led to optimum use of resources since they have generally been used to increase productivity based on imported inputs – protein supplements and fertilizers – while local resources, less productive lands and forests, have remained unexploited.

Norway banded together with Sweden and four other European countries that were not in the European Community to form the European Free Trade Area (EFTA) in 1960. Autonomy was thus preserved over agricultural policy and other issues that were considered strategic to their well-being.

Like Sweden, Norway has made use of various forms of policy and has been involved in an intense democratic struggle to formulate adequate solutions. Following World War 2 measures were taken to eliminate smaller producers and streamline production. In the 1960s the Small Holders Union protested and by the mid-1970s they were influential in establishing legislation favoring smaller, sometimes part time, operations. Surprisingly, production increased. Dairy and grain products accumulated and brought with them various related problems. There have been attempts to reduce dairy surpluses, impose fertilizer taxes to curtail input use as well as overproduction of grains and provide bonuses to farmers who reduced or stabilized production and established production quotas. Related research has contended that this mixture of policies has 'produced inconsistent outcomes and affected Norway's agricultural communities in highly variant and unanticipated ways' (Almaas, 1991:282). In some areas regional policies were applied. In the regions of best crop lands, for example, grain prices were subsidized to keep farmers out of the dairy business – which was encouraged in the more mountainous regions. Attempts to equalize farm incomes across the nation have been abandoned as ineffective. Rather there have been allocations to construct local non-agricultural industries in rural areas through municipally planned integrated rural development projects. Tension has increased between the Small Holders Union and the Norwegian Farmers Union which represents larger producers as they suggest conflicting measures for reduction of the surpluses.

Almaas (1991:228) recognises that the post-World War 2 policy established tendencies that persist in Norwegian agriculture:

- Conflict between large and small producers became organizationally institutionalized and resulted in the promotion of divergent policy goals;
- The northern (marginal agriculture) areas were depopulated;
- There was uneven development among regions;
- Water resources became polluted in regions of livestock production.

TEAM RIVALRY

Conflicts among categories of agricultural producers have become more common globally. These conflicts, while understandable, weaken the agricultural sector in general, even as farm populations are becoming smaller. The problems frequently center on topics related to exportation. Those producing for the export market favor policies different from those in the domestic market. Large producers have diverging suggestions for reducing surpluses. Small farmers have differing needs for credit. Livestock producers have needs for protein supplement imports with which others may disagree. A 1992 analysis of US farmers, conducted in the State of Ohio, compared opinions on several issues of the policies in effect between the local organizations of the Farm Bureau Federation and the National Farmers Union (see Table 2).

An example which further illustrates the differences between these two organizations is reflected in their positions on the

Table 2 *Attitudes to government policies*

	Farm Bureau	Farmers Union
Present agricultural policy in general	For	Against
International trade policy	For	Against
Capital gains tax policy	For	Split
Health care policy	Split	Against
Regulations	For	Split

Source: (The *Columbus Dispatch*, 30 January, 1992:2B)

General Agreement on Tariffs and Trade (GATT). The GATT involves more than 80 governments around the world and seeks to liberalize and expand trade by negotiating the reduction of trade barriers. Although the US was against the inclusion of agricultural products in the Agreement in the early years of its existence, as the costs of farm subsidies rose it became more expedient to seek global reductions than to continue to subsidize American farmers. The Farm Bureau Federation supports GATT proposals to open markets to free (untaxed) trade. The National Farmers Union, on the other hand, recognizes that most US farmers will gain less from free trade and that they will lose from the elimination of national subsidies. Obviously these two farm organizations represent different categories of agricultural producers and do not have the same goals for national production.

Similar positions of conflict on agricultural production also exist in other countries who participate in the GATT. Of course only to the extent that agricultural production is important in relation to other trade items included in the Agreement, are these positions more, or less, defended by each nation. France, an influential nation of the European Union, and with sizeable agricultural production, has called the GATT treatment of agricultural products to the attention of the group. It is problematic, in any case, for nations to jeopardize national agricultural production for more favorable treatment of other export goods, or for lower import commodity prices.

AGRICULTURAL INTENSIFICATION IN PORTUGAL

A final example of the creative use of agricultural policy comes from Portugal. Manuel Moreira (1991) reports that after the revolution to a more democratic form of government in 1974, the agricultural cooperatives of his country assumed many of the economic functions that had been held by the State. Cooperative regulations, left over from the time when coops were rigidly controlled by the Government, were changed to allow membership to small tenants. This opened new market possibilities to them, thus providing new sources of income and increasing the economic feasibility of small farms. Along with the opportunities for free unionism and other forms of democratic association came a shift from great (dictatorial) stability to great (democratic) instability. Agricultural interests varied in accord with the structures that had evolved over centuries. In the northern part of the country the predominant structure is the family farm which averages less than ten hectares (24.7 acres) of cultivated land and woodlot. This category of farmers produces 55 per cent of the Gross Domestic Agricultural Product (GDAP) (Moreira, 1991:299). In the South, large farms (100+ hectares) and extensive forms of agriculture predominate. They represent considerable wealth, but produce only 25 per cent of the GDAP.

Since the 1950s there has been a notable rural exodus from the northern part of the country. As alternative forms of employment were unavailable workers migrated to other European countries. By the time of the revolution the agricultural sector was very weak and the country depended on imports for much of its food supply. The revolutionary government, composed of new political parties and progressive ideas, encouraged the establishment of small and medium sized industry in rural areas of the northern region, a process that had been occurring with success since the 1930s. These industries were largely staffed by workers who spent most of their time farming. Many factories even adjusted work schedules to accommodate the seasonal needs of the region's agriculture. As industrialization increased, employment was available to avoid the rural exodus. Workers used their salaries to make substantial investments in their farm operations. Production increased. Factory managers report that it is to their advantage to have a workforce that is not totally dependent on salaries for income and the workers enjoy the freedom from total dependence on outside employment.

Moreira explains that the other side of the coin is the extensification of agriculture in the southern region which has largely offset the gains of the North. Agrarian reform appropriated many large farms to implant modern farming operations controlled by the workers themselves. These projects were for the most part unsuccessful and much of the land has been returned to the original owners whose management has resulted in its being even less productive than it was before the change of government. The innovative use of agricultural and industrial policies, as they have strengthened family agriculture and fortified the North, however, has been of great benefit to the nation and has prepared the way for entrance into the European Union.

COMMON AGRICULTURAL POLICY: THE CAP

The European Union has developed its policy orientation for agriculture over the years. Swann (1988:206) explains that it was an important topic for consideration in the Rome Treaty in 1958 when farming still occupied fifteen million persons, or 20 per cent of the working population of the Union nations. It was felt that the policy should:

- increase productivity
- ensure a fair standard of living for the agricultural community
- stabilize markets
- provide certainty of supplies
- ensure supplies to consumers at reasonable prices.

A 'fair standard of living' was planned through use of a series of common target prices (common throughout the Union after 1967) and intervention prices for agricultural commodities. Swann (1988:208) explains that a target price was established on the basis of what was fair in the region of most deficit of each product. Prices in other parts of the Union were established using the same price but subtracting transport costs to the deficit region. Intervention prices were set somewhat below the target price to signal the necessity for support purchases by the Union. The intervention price thus represented a minimum support price for producers. Imported products, if they arrived at the EU frontier for a price less than the internal price level, were subject to variable levies equal to the difference.

In terms of the other goals, productivity increased providing a bountiful 'certainty of supplies' and markets were more stable than in other developed countries as well as the world market. The goal of 'reasonable prices' to consumers could be questioned since world market prices were lower. It was felt, however, that importing at lower prices would jeopardize the other goals of the program.

While this system guaranteed income to EU farmers, it was neutral in terms of the type of production structure that was preferred. The result, as Swann (1988:216) admits, was that '...the CAP benefits the bigger and richer farmers most.' There have also been the tremendous problems of overproduction. The enormous surpluses of grain, dairy and other products have caused considerable strain on financial resources. Swann (1988:219) notes that 'Stocks of beef were so embarrassingly high in 1986 that the Commission had to hire refrigerated ships lying in Rotterdam harbour in order to store the surplus.' It became necessary to formulate reforms in the CAP, which took place in 1992 as cited above. These reforms have tended toward the idea of income supports and away from market subsidies as means of restraining production levels.

POLICY LESSONS

There are several points that we can learn from these examples in countries with varying historical experiences. Agricultural policy has great potential to bring the farm sectors along in step with other modifications in our rapidly changing nations. The relations among the farm, industrial and service sectors have traditionally received little conscious recognition as to their importance. A recent presentation on US Public Radio interviewed an American industrialist who decried the government's treatment of his sector as a separate, isolated entity. He explained that much US manufacturing is done in small firms which are disappearing at the rate of hundreds per year. The notion that the society is becoming service oriented, as opposed to agricultural, industrial and service, he criticized as being simplistic: 'All sectors are necessary for a sound society.' He cited the efforts of governments in other countries to stimulate cooperation among the sectors through their policies and the importance of inter-sectorial research. Possibilities of purposefully coordinating efforts and integrating planning among sectors represents an innovation that could be quite positive for agriculture and especially for family farmers.

We have seen from the countries reviewed that product price supports have had the effect of stimulating production, increasing surpluses, creating problems for exportation and raising land prices. Thus, while there have been clear benefits to farm income in general, the disadvantages are apparent. Larger farmers are disproportionately served by price supports and all farmers are encouraged to grow as fast as possible. In this situation there are many who will lose out and the future for all becomes uncertain. United States Senator, Rudy Boschwitz (1992) cites a study showing that the nations which offer the most agricultural subsidies also have the fastest declining agricultural sectors. Emphasis is given to short-term opportunities, crop years, export agreements, and the politicians responsible for policy formulation generally have most acute vision only through the next election. Agricultural development, though, is a long-term process involving major investments and careers. It deserves more rigorous consideration and adequate policy protection.

Income supplements, an alternative policy instrument that has been applied in some nations, provide security, but may delay decisions as to more efficient use of resources, land and capital as well as labor. In any case, strong nations, or agricultural sectors, or even family operations will not likely be built on any single policy orientation. Providing direct payments to farmers in exchange for the adoption of sustainable practices that also limit the production of excess food and fiber is an answer for countries with this type of problems. But nations that lack resources and need to stimulate production will not be able to apply this type of policy.

All this occurs within the policy process, which Knutson *et al* (1983:35 and 13) help us understand:

> The policy formulation process is itself often a key determinant of the content of policy. Thus, it is important to have an understanding of the nature of the policy process and the central actors who attempt to influence the direction and substance of policies for food and agriculture.
>
> Within government it is not unusual to have two or three separate economic studies of a controversial policy proposal. Such studies may be carried out from the perspective of a congressman, concerned about being reelected, from the perspective of the secretary of agriculture, concerned about its impact on farm income and US Department of Agriculture (USDA) spending, or from the perspective of the president's economic advisers, who are concerned about its impact on inflation and the budget.

It's time for innovative change. We have arrived at a point that is sufficiently critical for us to be able to recognize the necessity for more effective policy. Although known measures have the advantage of predictability, their long-term results are also predictably limited. We can now clearly say that *a major cause of the problem is the solutions that have been employed.*

It must be remembered that agricultural policy is influenced by beliefs and values in each society. Commodity groups and others take advantage of the family farm myth in the developed countries to stress the need for continuing traditional policies such as price supports. But these measures, as well as possibilities of increasing farm income through use of improved seeds, more fertilizers, or alternative production methods, appear for the time being to have reached their limits (Ruttan, 1988:55). To perpetuate traditional policies is nothing more than to whip the same, tired horse.

Policy makers in the governments of the developed countries are influenced by those groups who are most organized and best financed. They have also been impressed by the need for modern production organization. This has meant, for example, the 'vertical integration' of agricultural production. Harriet Friedmann, a Canadian professor who has conducted intensive research in this area provides the following description (1991:79) of the 'livestock/feed complex':

> ... the breeding and rearing of livestock has been transformed from extensive, handicraft husbandry techniques to intensive, scientifically managed continuous production systems. Livestock production has become more closely integrated with agri-food corporations both to buy scientifically designed, manufactured inputs, and to sell outputs standardized for mass markets and industrial food manufacturing.
>
> On the input side, livestock producers began to require feedstocks and many other technical goods and services. These were increasingly purchased from corporate feed manufacturers who design specific mixes of protein and caloric and other ingredients. The chain extends backwards to farmers producing specialized crops of soybeans and hybrid maize for sale to feed manufacturers. These farmers adopted capital-intensive production techniques and increased the demand for machinery. Monocultural farming necessitated demand for fertilizers and pesticides from the chemical industry. These products destroy nature's self-regulatory ability to maintain complex soils, plants and animals leading to ecologies less capable of reproducing their initial wealth of

resources. As such, these chemical inputs create conditions which demand their continued and expanded use.

This all developed after the words of caution from Peter Sinclair (1980:343) following his analysis of agricultural policy in several developed countries, more than ten years before:

> *Through vertical integration the corporation avoids investing in farm land, yet is assured of a market for the farm supplies produced by it and/or receives produce of guaranteed volume and quality for subsequent processing. The contracts usually leave the farmer with no more than routine day-to-day decision-making as the integrating firm contributes supplies, sets prices, determines key aspects of production schedules, provides technical supervision, etc.*

Sinclair concluded that:

> *... allowing for important regional differences, state policies either promote or do nothing to hinder the decline of commercial family farming. Small farmers understand this very well and have expressed their displeasure with state policy on many occasions.*

INTEGRATED RURAL DEVELOPMENT PROGRESS

There are other alternatives. A common thread that runs through the histories of the three countries reviewed, (Sweden, Norway and Portugal), is their use of *integrated rural development programs*. A strengthening of the rural community infrastructure: roads, schools, hospitals, commerce, can substitute, in many ways, the limitations of personal income. Local, accessible health care, appropriate education, and job opportunities that recognize the total life situations of workers all contribute to holding population in rural areas and fortifying family agriculture. Clearly such policy measures require cooperation among sectors in planning and programming to improve rural community life. Researchers in the US (Flora and Flora, 1988:126) have identified possibilities of community support:

> *There is an equally deep reservoir of good will toward rural communities as there is toward family farms in this nation. A substantial proportion of urbanites have rural backgrounds and many express a desire to live in smaller communities than their current urban residence.*

As seen in the countries reviewed, fulfilment of income needs may involve participation in different types of employment, at least for some family members and during part of the year. This diversification of income, however, provides a hedge against drastic changes in agricultural prices as has been the case in many countries during the last decade. Off-farm employment is not, of course, ideal. It poses many of the same problems that have been criticized in our discussion of urban occupations. But it needs to be a possibility. It shouldn't be necessary to give up farming because of problems that could be resolved if appropriate job opportunities were accessible. With more robust rural communities and lower living costs than found in the large urban areas, it is likely that there will be interest for location from the smaller, more flexible firms that are taking over the industrial scene from the corporate giants of the past.

Rural development represents a change of focus from traditional farm policy. It is not specific to certain crops, does not guarantee supplies of farm products as commercial brokers would like, will not aid the larger than family operations increase profits and may even raise consumer prices on some products. It will not, thus, be encouraged by many of the groups who currently influence farm policy in the developed countries. How can it be promoted? A rural development policy more adequate for families must be promoted by families. The examples of people who have organized to defend their beliefs and rights and life situations abound. Our review of countries has demonstrated that the most prevalent structure of agricultural production in the world is family agriculture. Yet in the developed countries the farm sector has come to be defended by other structures of production whose methods follow those of industrial organization and emphasize short-term profits. These methods require inventions and discoveries to reduce risk: genetically modified plants and animals, production integrated with a flow of inputs from the urban area. In return they guarantee a constant supply of uniform quality product for subsequent processing. These interests aren't necessarily the same as will most benefit families who have chosen to dedicate their careers and fortunes to agriculture.

An example of divergence is that of soil and water conservation. Families in all parts of the world have traditionally valued the construction and maintenance of terraces and irrigation ditches as well as crop rotation, wind breaks, farm ponds and other practices that assure resources for future generations. But these measures are, according to the new defenders of modern agriculture, time consuming and uneconomic. Many of us have heard these ideas and mulled them over to the point of seeing their merit. We may even have come to consider ourselves as commercial producers first and family producers by coincidence. Irritation with environmental protectionists symbolizes this new orientation. This poses a quandary. The identification of where our real allegiance lies will determine our interests in defending one or another form of agricultural policy. Farmers surveyed in the county of Devon, southwest England, by the Centre for Rural Economy expressed varying opinions when asked about governmental controls on pollution (Ward, 1993:3):

> ... those who most strongly approve of the strict regulatory approach to farm pollution tend to run more diversified businesses and show a low level of commitment to family continuity. On the other hand, those who most strongly disapprove of the regulations and feel that the pollution problem is 'over-hyped' tend to be traditional owner-occupied family farms committed to succession.

Thus we can understand the problem as involving values, the topic of discussion in Chapter 6. Government controls pose a possible threat to the future that we would hope for for our children. They may alter patterns of investment, change long-standing family goals and modify the future lives of cherished people on cherished land. But perhaps it is time to reconsider some of our traditional values. Maybe we should remember those of the family who have moved off to the city as well.... Environmental protection is not a scheme to castigate farmers but a hope for a better future for ourselves and our descendants. My hope is that the reader who has reached this far in a discussion of family agriculture has enough determination to fight on for its preservation even as that future may not be a repetition of our past traditions.

Does this mean that we need to increase our cooperation with the environmentalists? Friedmann's comments earlier suggested that some developments in agriculture have reached a point that the aims of environmental protection are actually more compatible with goals, especially long-term goals, of families than some aspects of modern agriculture itself. The 1992 analysis of the US situation by Browne et al (1992: 192) reinforces the point:

> Improving environmental quality and conserving resources in rural America may involve new technologies, but they are not absolutely necessary for solving our problems. Techniques

that come closer to maintaining sustainable agricultural systems already exist. Ironically, many of these techniques would be more profitable to adopt without current farm programs. We need to recognize the many factors causing environmental degradation. The solutions will require compromise and cooperation among various competing social interests. Public policy, as opposed to market solutions or voluntary adoption, is needed to bring the best possible chance for a sustainable U.S. agriculture.

An example comes from a grassroots conservation program in Australia in which various sectors with interest in agriculture have come together to resolve problems. Pretty and Howes (1993:44) suggest that there are possibilities of overcoming past differences:

Although there are many factors accounting for the success of the Landcare programme, one of the most significant is that both environmental and farming lobbies worked together in partnership to establish the programme, leaving behind entrenched positions that had prevented earlier collaboration.

FARM POLICY ISSUES

With families taking the initiative in the struggle for more adequate agricultural policy, what are some issues, consistent with rural development, that need to be examined?

Financial programs to guarantee timely credit have been the tradition of the US Farmers' Home Administration. Its diverse programs to serve varying individual needs, grants and loans to rural communities, loans for business and industry located in rural areas and housing improvements represent the type of program that would be formulated to carry out the 'adequate' farm policy here advocated (Knutson *et al*, 1983):

1. Rural electrification, while extensively established in the developed countries, has great potential for modernizing the agriculture of other areas and improving the well-being of farm families.
2. Crop insurance could be financed in ways that concentrate on protection rather than guarantees of profit and increases in land value. Here Browne and colleagues (1992:74) suggest the possibility of 'area-revenue insurance.' Farmers in similar agricultural regions would pay at the same rates for an agreed upon level of coverage, and receive in accord with the

degree that area revenues for a specific crop are below average. Such insurance would be less expensive to administer than individualized programs and could have the advantage of promoting farmer organization for decision making. It would also have a stabilizing effect that would help farmers manage risk.

3. Educational programs by agricultural extension services and agricultural universities could help producers manage the disproportionate levels of risk that they face. The levels of capital, linkages and opportunities are different for family operations than for larger producers and require specialized programs.

These are only a few of many possible ways that policy could be better for families if they were more involved in its formulation. This is not to say that the organization of farmers to defend what we all feel should be freely offered, will be easy. Sinclair (1980:346), viewing the American scene, wrote:

> *The family farmers that survive will continue to be difficult to mobilize for change because their dependent class position is countered by a work situation which allows illusions of entrepreneurial independence to survive.*

We still see ourselves more as owners and employers and suppliers than renters and hired workers.

Finally, what do the authors who have contributed to our policy discussion have to offer for our future?

> *... it is also possible that the artificial harmony of corporatist decision-making will break down and spawn more radical farmers' union movements to press for fundamental change.*
>
> *(Sinclair, 1980:346)*

> *Given social changes, farms depend more on rural communities than rural communities depend on farms. These new linkages between farms and rural communities intensify the need to diversify job opportunities and rural development.*
>
> *(Browne et al, 1992:72)*

> *To arrive at a cohesive policy position, compromise among the members of a group with respect to goals, values, or beliefs is frequently necessary. The willingness of the members to compromise is a source of strength. Without compromise, constant friction among members of the group is possible. Sufficient friction results in an inability of the group to arrive*

at a policy position. This is particularly a problem among farm organizations where major differences in goals, values, and beliefs frequently exist.

(Knutson et al, 1983:11)

... if farm and food workers throughout the food regime joined popular movements for healthy food, democratic land use and environmental protection, we could begin to plan agriculture and food production appropriate to local ecologies and diets. However challenging, this is a democratic, and stable, alternative.

(Friedmann, 1991:89)

10

An Enemy Within

We have examined family agriculture as a production structure with a distinct social role in the societies of various parts of the world. The review of physical aspects, of ecology and sustainability, provided insight as to how the family structure relates to its resources. We have also seen how agriculture is directed in various ways by the national policy that is established in each country.

This brings us to the most critical component of our analysis, the producer. In the middle of tremendous change, in terms of difficult economic conditions and complex physical relationships, but also in terms of alternative opportunities in urban areas, the present global farm population has made decisions that have brought more change to the sector than has been seen in many generations. Some have expanded farm size, others have taken off-farm jobs to increase income and others have left farming altogether. These modifications have required substantial restructuring of a psychological nature in every one of the individuals involved. On the surface the change appears to have resulted from external conditions. Dialectical logic, however, has taught us that change is generated by *internal contradiction*; it may be within an apple, an individual or even a government. Something inside is against the circumstances of the present situation. There is dissatisfaction. This sentiment is fed, in the case of individuals, by the information that is received. Analysing the US situation, Browne and his colleagues (1992:17) explain that the 'rural America of our dreams persists ... wrapped up in our desire for ties to the land, economic independence and community support.' But there are other, conflicting goals. On a collective basis it has been said that our love of growth has led us to neglect values that were previously of importance. Material investments and goods take on altered relevance, out of proportion to our traditional needs. Larger farms suggest the desire for increased personal stability and higher standing in the community. Automobiles are transformed from means of transportation to status symbols. Even clothes, which have been used for status for many years, assume new aspects as it becomes important not to be seen too frequently in the same outfit. These examples suggest the necessity for more cash income. With increased resources we would be able to participate in the 'improvements' – of the farm, the garage, or at least the wardrobe. Suddenly the values that have been relied upon to provide feedback as to 'how we are doing' become outmoded. Things don't seem right anymore. We sense an internal contradiction.

THE OHIO VALLEY SYNDROME

Bob Jacobson, US dairy extension economist at The Ohio State University, often speaks of the 'Ohio Valley Syndrome.' He has observed that even fourth and fifth generation dairy farmers within 75 miles of the Ohio River in Ohio, West Virginia and Kentucky, seem to be running down. Production per cow is declining, or at least is not increasing, and the producers themselves are becoming marginalized from the innovative practices and methods that are coming into use among other, modern dairy farmers. Talking with some producers in the area reveals that they feel they've been pushed around by the government. There are those, on the other hand, who state that they couldn't continue without the 'checks from the government.' At the same time they dislike the implication that they're 'on the dole,' especially since they work so hard. These are internal contradictions that change the way we feel about what we do. Some producers express the desire for a better life for their children. It is thus not surprising that their children show little interest in continuing with cattle farming. Some are looking for a new game. Other vocations appear more interesting. An important part of this process is the social communication that circulates within families and communities, and between rural and urban areas. Any perception of a reduction of status associated with farming as a profession generates questions within us.

    There have also been effects of the powerful trend toward farm mechanization in the developed countries that function as a contradiction to the well-being of agriculture as a sector of society and have driven many families from the land. B F Stanton (1989:15 and 22) relates that in the US:

*Average farm size began to increase after 1920 as tractor
power began increasingly to replace horses. The great leaps
forward occurred between 1950 and 1969, at the same time as
farm numbers were cut in half, another indication that this
was the period of greatest structural change in U.S. history.... By 1982, 85.6 per cent of the land was in farms with
260 or more acres. The proportion of total agricultural land
farmed in units of 1000 acres or more has increased steadily
across the twentieth century.*

It became not only socially acceptable to leave agriculture, but a
means of bettering one's situation. Without doubt it was necessary that the US farm population be reduced to strengthen the
incomes of those who remained. This is not to say that the trend
should continue unmonitored. There are also changes in off-farm
opportunity, costs of urban living in relation to salary levels,
quality of services in rural areas and other factors associated with
'rural versus urban' decision making. What may have been a
plausible decision a few years ago could be foolhardy today.

Bill Heffernan, a rural sociologist at the University of
Missouri, has analysed American farm families who left
agriculture. He recognized (1986) another contradiction in the
research that has been conducted to 'improve' agriculture:

*Most farm families did not voluntarily leave their farms. They
were forced to leave when they could not survive on their
farms because research designed to increase labor efficiency
led to over production and lower prices.*

He continues, to relate the reaction to these problems:

*... a recent study done by the National Institute for Occu-
pational Safety and Health reports that farming ranks in the
upper 10 per cent of 130 high stress occupations. Stress occurs
when one is unable to control events which have a serious
consequence for one's life. Although this is a general
characteristic of all farmers, family stress is greatly increased
when a farm family faces foreclosure. Because the family's
life is so closely tied to the farming operation, an occupational
change usually leads to a change in the family's way of life. It
also usually means a severing of social ties as the family is
forced to move to other communities in search of employment.
The personal cost to family members can not be calculated in
economic terms.*

Within the individual there is perplexity. Jenny Cornelius teaches
courses in stress management in rural communities in England.

She has found (1993:2) that '... social conditioning over many generations has resulted in the farmer feeling he must be the provider for family and staff; be strong; and cope in diversity without complaint or show of emotion.' Stress is an internal contradiction reflecting feelings toward one's conditions of life. To the extent that it is recognized as a *response* to a social situation – not created by the individual – it can be dealt with. Stress is, after all, associated with the attempt to act responsibly in the face of difficulty. Irresponsible people are likely to experience less stress. However, for those who don't achieve this understanding, says Heffernan (1986:203), the result can be drastic:

> *Those unable to cope with stress may ultimately commit suicide. In Missouri we have experienced almost a fourfold increase in suicide among farmers less than 65 years of age, from six suicides in 1982 to 23 in 1983. As the economic crisis facing Missouri farmers intensifies, health authorities anticipate even greater increases.*

Pretty and Howes (1993:6) report from England the statistics that explain the need for the courses given by Cornelius:

> *According to the Office of Population and Census Surveys, farmers and farm workers are about twice as likely to commit suicide than the rest of the population, and suicide is the second most common form of death for male farmers aged 15 to 44 years. Farmers are increasingly recognised as suffering ... stress and deteriorating confidence....*

The farmer who commits suicide feels that the full weight of the problem is on his own shoulders. He doesn't recognize the relation that his part of the problem has with other parts. He feels that he is alone – he should have learned something about dialectics.

The reduced number of farm families, in turn, becomes a contradiction in the rural community (Heffernan, 1986:203):

> *Rural trade centers, which draw a major portion of their economic support from farming, have shared the farmers' economic conditions. Fewer farm families has meant the need for fewer goods and services in rural communities. The uprooting of farm families has carried over to include the families of small business persons, as small towns across the country have seen their economic base decline.*

THERAPY FOR INTERNAL CONTRADICTION

Internal contradiction is the seed of change. It isn't in itself good or bad. Nor can it be decided that it should or should not exist. It *does* exist and generates change. This is not to say that we must let change take its course in an undirected manner. To take charge of our farm futures, and family agriculture in general, we must react.

First we need to identify the problem or problems that have created the contradiction. It is, as we have learned, the result of a series of 'quantitative' variations and may take the form of changes of attitude related to personal and socioeconomic occurrences. These changes may be positive, in the form of alternative occupational opportunities or lifestyles, for example, or they may reflect the problems of farming operations and/or rural life. A contradiction for many families is the inheritance to be received by siblings and offspring. When there is more than one heir to parents' farms the difficulty of a just division without selling the farm, especially in these times of high land prices and loan problems, is very real. At the same time, from the parents' point of view, the possibilities of dividing one's property among children so that there is opportunity for all, while ensuring the continued existence of the family farm, can be very complicated. Through family discussion – which is sometimes neglected 'to avoid making it worse...' – we come to understand personal attitudes more fully.

To formulate a basis for action on the inheritance example given, or other contradictions that arise, it is necessary to identify alternatives. While the pasture often looks greener on the neighbor's side of the fence, a rigorous examination of alternatives may reveal details that clarify their desirability. It's not uncommon to discover that things are not as bad as first anticipated and that's great for alleviating contradiction. Alternatives can be sought or they can be created. The diversification of farm income, for example, could reduce dependence in certain areas. By creating our own opportunities we increase personal satisfaction with the results.

Changes in agriculture today are rapid and drastic. Many of the contradictions are not internal to the individual, but internal to a region, to producers of certain crops or to some level of government. Possibilities of moving commodities over vast distances, disturbing local markets and creating new frontiers of competition, only to disappear when profit margins of the merchants are reduced, need to be taken into account by local producers. The stability of traditional family production, in spite of short-term agitation in the market, should be valued at the government level, and recognized by individual producers. On the other hand, the use of industrial methods, when applicable to agricultural production, is probably here to stay in countries where labor costs are high. Broiler production is a prime example. The methods of mass production have greatly reduced costs per bird, making it difficult for producers who use more labor intensive methods to compete in the market. We need to keep up with the new ideas, evaluate them with the help of information services, field days and personal examination, and adopt those that are feasible. But when a contradiction is internal to a region, for example, then it affects all producers of the region and will require their collective response for a satisfactory solution.

Agricultural producers in the developed countries today are the survivors of a process that has eliminated the majority, over the past few decades. Entrepreneurial ability, good business sense, is more important than it has been in other times and in other places. Not everyone has this ability or is interested in developing it: for them the contradictions have been more significant. There is much to be said for those who have made it to the present time. It is that survival spirit that will carry us beyond this and future crises.

BANANAS AT THE AIC – 1954

The American Institute of Cooperation held its annual meeting at Cornell University in Ithaca, New York, in 1954. Meeting people from 36 states at the social activities of the first evening was a first for me and even more exciting than having won a high school scholarship to the event. At sessions in the following days we learned of the importance and benefits of cooperation. There were catchy slogans that are remembered over the years – the banana that leaves the bunch, gets peeled! It seemed certain at that time that within 20 or 30 years the American farm population would be totally organized and united in defending its interests in the increasingly urban population. American farmers have historically organized movements to influence local, state and even federal government policies that affect their professions and their lives. This is another tactic for dealing with internal contradiction in organizations and institutions. As has been mentioned several times during our discussion, organization is essential.

Today, as we have seen, attitudes have become rigid concerning cooperatives. There are those who remain convinced that they are the salvation of the farmers. Others have become discouraged and show little interest in participating. For the majority, cooperativism has become a fixed concept for which opinions are established and are not easily changed. But this is different from believing that there is no future for cooperation. Of course, most of the current problems faced by rural populations throughout the world cannot be solved by individual action – there is a need for innovative ideas as to how producers can join together to exert collective force. In democratic societies this still means 'grassroots movements.' While we tend to pay more attention to national news and can more readily recite the names of our national leaders than those of our municipal authorities, we are actually mere spectators of the national political scene. As individuals, we can no more alter the activities of national politics than we can the results in World Cup Soccer matches. Democracy grows, as Robert Chambers mentioned earlier, from the bottom up. Three or four people who sense a contradiction, work well together and can bring the specifics of a problem into focus for others who are also affected, may constitute an effective nucleus. Leaders who have organizational and planning abilities are critical. There is nothing new in emphasizing their role for the application of organization building techniques. But the other participants too are of critical importance. Often the unifying process of seeking change is as useful for them as the realization of change itself.

Commitment increases as initial participants recruit others in attempts to involve a substantial portion of those affected by the target problem. In the developed countries of recent years many farm leaders have sensed increasing difficulty in motivating their colleagues to come out to meetings and participate. Larger work loads – sometimes in salaried positions – divide our interests and sap our energies from any activities not seen as essential. There is also a feeling of helplessness, the fatalistic idea that so little can be accomplished it's not worth the time and effort. Still, people would surely not agree with this position when it is carried to the point of survival in their professional lives. Individual force is limited. Sometimes we have actually seen justice in colleagues being forced to leave agriculture, feeling that they were not strong enough, or independent enough to survive. But the same

socioeconomic conditions that eliminate some will also be
problematic for others.... We must organize.

IDENTIFYING OPPORTUNITIES

It is a necessary characteristic of those hoping to initiate change
that they be able to recognize the readiness of others who share
the same problematic conditions, to react. There also is the need
to identify possibilities for success. Attacking well-chosen
problems, or breaking problems down into a series of steps that
can probably be solved generates enthusiasm among participants
to overcome obstacles and the costs involved in participation.
Most problems actually constitute opportunities to strengthen
action groups.

In countries that are well on in the processes of development,
those in Latin America, for instance, there is reluctance to come
out to evening meetings that is directly correlated with television
programming. Sometimes it is the nightly 'soap opera' type
programs that hold high levels of attraction for men, women and
children, in contrast to the general political interest and awareness that have been traditional in these countries. Heated
discussions are still a characteristic of coffee shops and little bars
where stronger drinks are sold. But the development of the mass
media has been associated, all over the world, with growing
apathy for organized group activity and an increase in isolation.

Organizations don't need to be formal, with dues and fixed
requirements. They can arise spontaneously in response to a
particular contradiction, and dissolve when the participants are
satisfied that the necessary actions have been taken. On the other
hand, groups which began in reaction to a particular problem
may grow from one problem to another, becoming more
permanent. In view of the dynamism of modern agriculture and
the rapidly changing situations of producers, less requirements
on group participation may well encourage higher levels of
cooperation.

THE 'ONE VILLAGE, ONE PRODUCT' MOVEMENT

Following World War 2 Japan elaborated a series of national
development plans which included directions for progress in
agriculture. Some plans stressed modernization, others production increases, as would be expected. Then communities began
working together to create 'isson ippin' (a 'one village, one

product' movement). To a western observer this probably suggests village specialization – one village raising rice, another beans. But closer reading reveals the 'one product' to be 'community spirit, pride.' It is explained by Isao Fujimoto (1992:12) that the goals of this plan stress the exercise of local control, building community spirit and providing the outside world with products of a 'distinct local flavor.' One Village, One Product started as a grassroots movement in reaction to the feeling that '... with more government aid and administrative presence, communities became more dependent.' Soon the government began giving encouragement. The movement recognized differences between economic growth and development, which was seen as a more worthy objective. Community organization was cited as being advantageous to the movement, to encourage local pride and commitment. This was not to recommend isolation, rather it was to develop the people's 'own identity as a vital component to a region.' The result would be a complement to all (even global) development.

In a conference with farmers, Fujimoto displayed buttons with various slogans on them. He asked participants which ones were most important for them. The slogan most frequently chosen read: 'When the people lead, the leaders will follow.' The decision had been made to concentrate less on producing (separately) and more on living (together).

It is good news that internal contradictions do not need to be left to run their course. They need not mechanically evolve and generate change. Rather their recognition creates in us an awareness of related factors and allows us to put the contradiction in its context. This informed awareness generates the consciousness that is needed to take stock of the situation and react.

What constitutes a contradiction for some producers, at their level of awareness, may not be the case for others. Japanese producers who have migrated to Brazil complain that the 'intermediários,' often just persons with a truck at their disposition, purchase agricultural products on farms for a fraction of their market value, taking advantage of the producers' lack of knowledge of market prices. Agricultural leaders in São Paulo feel that they need to participate in the transformation of their relations with the market structure. Levels of consciousness vary with our positions in society and the information that we have – and these factors are associated with the contradictions that we sense.

David Vail, writing from Sweden, cites a characterization of Americans as 'utterly absorbed in the present.' A commonly used example is that they no longer even associate the time of day with its hourly context: rather than the usual 'quarter to six,' by the new digital timepieces it's '5:44,' or '3:17,' or '10:22,' – times that were previously unnoticed within their general position on the clock. Vail (1991:260) explains hopefully that '... most Swedes seem to have a sense of history, of connectedness to the past and the future. This is manifest in their willingness to pay for the preservation of family farms and open landscapes.'

Whether the contradictions that we sense involve personal farm enterprises, the direction of agricultural research, commodity programs in state farm policy or a modification in the cooperative milk pick-up route, it's through the consciousness of our past, an awareness of the present, and aspirations formulated for the future that we are able to deal with them and move on to tomorrow.

11

Information Systems and Survival Techniques

Maria de Sousa rinsed her hands after morning chores, returned to the house and switched on the 'unidade de comunicação' or 'UC' (communications unit) that her family had received from the municipal government. Since they had no neighbors within two kilometers the unit was installed in their home, powered by solar cells mounted on a pole outside the door. It had been a jump from no outside communications, to the unit. There is still no mail service and the standard telephone lines stretch less than five kilometers from the town. It is 2005 and Maria is grateful for the 'animation' that her family has received from the UC. Being connected to a digital radio she can use the UC like a telephone, to call relatives in town, but now she is interested in checking the neighborhood bulletin board. She types in her password, logs into the municipal computer and clicks on the 'Bulletins' icon. There are many messages listed on the screen, one could just read all day: several neighbors want to join together to hire a truck to bring fertilizer from São Paulo; the local livestock trucker is free next Tuesday if anyone has cattle ready for market; there is a sale at a clothing store in town.... She checks the price of cattle to confer with her husband and glances at her horoscope for the day. As she goes back outside, the UC stays on in the screen saver mode. Her mother always complains that it is a waste of energy, but the extensionist explained that on 'idle' the computer uses nearly no energy, and that when it is used several times during the day it is less wearing to leave it on than to 'boot up' each time. Anyway her mother really doesn't understand the modern technology. She had once said that they'd never have television and now they've had it for nearly twenty years!

A situation that has caused problems for farmers and other rural residents for generations is the isolation and sparse communication with distant neighbors. While the scene described above may sound like science fiction in the Third World, it actually hasn't been that long since personal computers were considered toys even in the developed countries. Maria's UC may well become a reality, as similar technology is becoming common in some parts of the world. It is another example of how family agriculture can be strengthened. Networking services are coming into use for financial and business concerns and there are also innovators in the agricultural sector. In *Rural America at the Crossroads: Networking for the Future*, published by the US Congress, Office of Technology Assessment (1991:9), instructions are given for the organization of Rural Area Networks (RANs) to link rural businesses, schools, health providers, local government offices and others interested in improved communications. The benefits listed include:

- diffusion of advanced techniques;
- fostering of cooperation and community ties:
- increases in leverage in the marketplace;
- competition with established communications media to improve service to meet community needs;
- linkage to communications satellites for a wider range of services.

Since the networks are generally based on radio contact their cost depends more on total demand than on population density. Digital radios link with personal computers and telephones. The technology is available, but how these mechanisms are used for development in rural areas depends upon those involved. In the hypothetical example given above, the municipal (county) government has taken the initiative to install a single central computer and communications system, distribute the remote units and coordinate their use. Expenses are initially covered in the municipal budget and bills are sent (electronically, no paper is necessary) to the commercial users' accounts. The local agricultural extension agents conduct educational programs on Communications Technology and provide individual assistance in use of the UCs. They also make abundant use of the network to reach their clientele.

Messages can be sent to any network users. This allows Maria to communicate, in writing, with friends, storekeepers and city officials. Maria is finally convinced that she did not waste her time learning to read and write, although she had used these skills rarely before receiving the UC. Since there is commercial information on the bulletin board the storekeepers pay a part of the costs and Maria's family pays nothing. There is also no wasted paper from sales bills and advertisements to consume resources and pollute the community. Neighborhood meetings can be organized quickly and easily by opening a discussion on the bulletin board and supplementing it with messages to individual addresses as necessary. But this is just the beginning of the advantages! Although neither Maria nor her husband ever had a checking account, now they can pay bills at the same time that they 'call' the bank for the balance on their savings account.

Groups of farmers are able to join together to sell quantities of produce for more profit and negotiate larger purchases at lower prices and with advantageous delivery arrangements. Machinery parts can be ordered with less confusion and more speed and all that is done is automatically saved for future reference.

Opinions on topics of interest can be formulated at ease, typed on the screen and corrected as necessary before the 'send' command is given. The result is that individuals participate in networking who would not risk an opinion in a public meeting. When colleagues do come together their positions on the topic are already known so that communication depends less on personal speaking ability to express ideas. Those who have incompatible work schedules can still participate during the time that they have, in the bulletin board discussions.

It was a major breakthrough when libraries of the communities, the states, then of the nations and now of many parts of the world were united in a common call system. With this integration the same steps are taken to electronically review the holdings of a foreign library as for those of the local institution. Being able to access these directories by calling into the library from home or office has been a tremendous boost to those seeking information. This includes kids doing homework. While it may seem to be an elaborate system for personal, sometimes incidental use, it's through the pooling of the total public and private demand that the system can be justified in isolated areas. For these reasons it's an exciting idea for family agriculture.

NETWORKING FOR HEALTH AND EDUCATION

Networking is also useful to upgrade the services of community institutions and agencies. It is possible, for example, for small hospitals to maintain their competence even in the face of enormous innovation in the health care services in larger, urban areas. Doctors can monitor and record patients' conditions, initially without their leaving home. A second opinion does not require having a doctor physically present. Consultation can be made by the physician to research institutions in the larger cities for up-to-date treatment procedures. They can also receive a regular flow of medical instruction. The American KARENET System in Texas (cited in the *Crossroads* publication) was established in 1985 and has proved its usefulness in serving small rural hospitals as well as reducing travel problems of patients.

The 'Big Sky Telegraph' established by Western Montana College has demonstrated the usefulness of a bulletin board network system. The 'Telegraph' was first used to link Montana's 114 one-room schools to each other and to the College. Since its establishment in 1988 it has progressed from being a resource support system for the schools, to provide recertification programs for the teachers, general educational programming, service to businesses and to individual users. *Crossroads* reports that:

> *About 100 community sites (including schools, libraries, county extension offices, chambers of commerce, women's centers and hospitals) will be equipped with a modem to connect their computers to Big Sky's network. Circuit riders travel throughout the State to introduce people to the technology and familiarize them with its offerings, and local system operators are given training to help the community use the services. . . . The Telegraph is a tool for enhancing education, for broadening and strengthening community, for facilitating economic development and for building grass-roots democracy.*

Although our discussion has been chiefly in terms of written or spoken communication, Big Sky also has the visual component by which the computer monitor serves as a television screen to receive films and programs including courses from distant institutions. The system has served as a model for the development of similar networks in other states of the United States.

The *Crossroads* publication also warns, however, that the new communications technology could possibly do more harm than good. The problems of rural areas still include poverty, high drop-out rates from school, poor health care conditions, inadequate infrastructures and a shortage of capital. As large firms are able to contact local people directly they may well undercut local businessmen and undermine community institutions. There is a danger that those in isolated regions, when learning of opportunities in other areas, may be pulled in by the glitter of communication centers. But the technology itself is valueless. It is a tool that has potential for many uses. Which of these are selected could be the responsibility of the community base that is using the information. Regulation will be necessary and should reflect the values and interests, first of the users, and then of the larger society.

In terms of personal experience, having subscribed to a network for some time, there is a high degree of participation and democratic action. At times when issues are raised that attract wide attention there will be over 100 messages in a day. These tend to be short, an average of ten lines, and from all over the world. The subjects posted in the list of messages received (like letters in a mailbox) usually make it possible to decide whether each one would be worth reading. Otherwise the 'delete' option is a fast route to the trash.

HOW DO WE USE COMMUNICATION TECHNOLOGY?

Each group with an interest in networking will need to develop a plan as to how the communication technology will be used. This plan may require revision periodically. In the network of my experience a rash of jokes on the bulletin board resulted in a number of complaints and eventually the split off of a part of the group that was interested in 'strictly business.' When a system is being planned considerable negotiation is necessary. This is a positive process as various groups and factions of the community come together to share the risks, the benefits and the costs of the service. And like the 'party lines,' characteristic of the telephone service that stretched across rural areas in the early part of the century, it generates a lot of interest too. Communities must decide who will participate, what services will be offered, how costs will be divided, and much more. The *Crossroads* publication makes the following suggestions for success in community systems in the US:

● The technology is new. Vision, imagination, ingenuity, and enlightened leadership are needed.
● Programs and policies must be concerned with a holistic orientation.
● Programs must be efficient and effective because of very limited funding.

● programs need flexibility to serve diverse interests.
● National development policy must provide technical assistance and education to communities to familiarize them and assist them in planning and devising communication based development strategies.

Interest in the US for the application of communication technology in rural areas has, in these early stages, been focused on possibilities for community development and economic growth. The new technology is not a tonic that can cure long festering social and economic wounds, but it does represent an innovation that could make new links possible among the sectors of the economy, reduce some of the longstanding problems and help rejuvenate disadvantaged areas. The *Crossroads* publication on which our discussion is based was elaborated at the request of a US Congressional Committee. The first pages begin:

> *Since 1970, the U.S. trade position has steadily worsened, while those of our major competitors continue to improve. Much of the increased trade competition is in the area of primary goods and low-tec industries – the industries in which rural areas have traditionally specialized. Rural areas can contribute to an improvement in the U.S. trade balance if economic development in these areas leads to greater economic diversification and/or a shift to those industries – such as services and high-tec manufacturing – that are growing in demand worldwide.*

This passage is followed by suggestions that each community inventory its resources (both natural and human), evaluate its interests and begin making decisions. In the case that human resources are adequate, but jobs are lacking, for example, a development program should address ways to increase local employment. The results may include development strategies that will optimize the use of communication technology and bring various groups together in an effort that could be beneficial for everyone.

While development of the community is not the highest priority of our discussion, family agriculturalists are certainly included among those who will be benefited by its success. And the advantages of improved communications apply directly to farmers too. Vernon Ruttan (1991:228) at the University of Minnesota, in speaking of the future, has stated:

> *The means of achieving greater yields will be extremely sensitive to new knowledge and information. If they are to be*

*fully realized, research and technology transfer and the
information and management technology to go with it must
become increasingly important sources of developments in
crop and animal productivity.*

NEW FORMS OF COMMUNITY ACTION

There are many new ideas and practices coming into use in rural
areas. Unfortunately many of them are never heard of beyond
their local communities. A few years ago a graduate veterinarian
in rural Brazil returned from the university to his hometown.
Many of his colleagues were unemployed and he hadn't come up
with any options. While giving his father a hand on the family
farm he began to make contacts with neighbors who worked with
dairy cattle in the area. Within a year he had established a
program through which 20 farmers received a monthly visit each
and the group held a meeting once a month at which he presented
technical information on a common problem and led a discussion. They each paid one 'minimum salary' (approximately fifty
US dollars) per month and the recent graduate was earning more
than many of his university professors! This type of innovative
action can be enhanced even more through use of communication technology.

There is also more formalized activity at the community level.
We shall now examine some formal organizations that have
evolved to improve community conditions in the areas of land,
labor and capital.

The Community Development Corporation

Community Development Corporations (CDCs) were first
organized in the US in the 1960s to solve the problems of
debilitated communities and residents without opportunities.
Robert Zdenek (1987:113 and 115), President of the National
Congress for Community Economic Development, explains the
concept:

> *The CDC emerged ... as a vehicle for a comprehensive
> strategy of business, social, and physical development,
> designed to generate both human and financial capital. It was
> conceived as an institutional base for revitalizing the com-
> munity. This comprehensive role indicates that CDCs are
> inherently suited to affect the overall coordination of the
> various new forms of economic organizations and new
> strategies for community development.*

CDCs ... engage in a wide range of activities: housing development, commercial revitalization, business financing and assistance, daycare centers, job training and placement, social service delivery, cultural activities, advocacy, and the creation of other institutions such as credit unions, co-ops, and loan funds.

The Mountain Association for Community Economic Development (MACED) in eastern Kentucky was able to organize 90 small rural banks to offer loans at reduced rates of interest to low and middle income residents. It also founded the Rockcastle Lumber Company to organize small independent sawmills, buying and stockpiling lumber and selling it in the national market at the appropriate time (Zdenek, 1987:115 and 120).

Having personally worked for some time in the Hough Area Development Corporation, a CDC in Cleveland, Ohio, I am confident in saying that residents develop new hope and great enthusiasm when they find ways to participate in lasting solutions to community problems and create a structure for alleviating individual and family problems as well. As Zdenek (1987:122) summarizes: '... it is critical to view community economic development as a long-term process that fuses community control and direction with specific tools and outcomes.'

The CDC furnishes the organizational structure for a number of other survival techniques, including those which follow.

Community Land Trusts

On the question of accessibility to land, one possibility is the CLT. White and Matthei (1987:52), from the Institute for Community Economics in Greenfield, Massachusetts, explain:

The CLT acquires land through purchase or donation, with the intention of continuing to hold title, thus removing the land from the speculative market. Where specific uses (such as housing) have not already been established, appropriate uses of the land are determined by the board or a committee of the CLT. Parcels of land are then allocated to individuals, families, cooperatives, or community organizations, through long-term, renewable, inheritable leases that provide security similar to that provided by traditional ownership. When a leaseholder ceases to use the land, however, the lease is terminated so that the land can be reallocated to others who will use it. Thus absentee control of land is prevented and the community can provide access to land for future residents.

Leaseholders pay a fee for their use of the land, but they need not finance an acquisition cost in order to obtain access to it. Lease fees are used by the CLT to acquire additional land, pay property taxes, and cover its own administrative costs.

Though leaseholders cannot own the land, they can own housing and other improvements on the land. When a lease is terminated, the leaseholder can sell the improvements, thus retaining the value invested in them.

The possibilities of land trusts have yet to be explored fully, but *The Community Land Trust Handbook*, written by the above authors and several others, provides an interesting example from Marin County, California, where a group of farmers and 'ranchers' had a strong desire to continue, and for their children to have the opportunity to continue, with their lands for agricultural purposes. Located just outside of San Francisco, there was intense pressure from speculators interested in developing housing projects and ranchettes as urban getaways. The area, which includes approximately 130,000 acres, was producing 25 per cent of the fresh milk consumed in the city. There were many third and fourth generation producers who used modern techniques and were reluctant to see their farms destroyed in the spread of urbanism. Without some form of organized reaction to the spectators the land base would have been gradually reduced to the point that a critical nucleus for production would disappear and the farm community would be lost.

Ellen Strauss is one of the farmers in the region who worked for several years for a community solution. She considered the benefits and restrictions of agricultural zoning, conservation easements and other legal forms from the point of view of individual rights, but sought to defend community goals as well. Agricultural zoning would have restricted those who wanted the option of being able to sell off some frontage to raise capital, or to retire. On the other hand, a land trust, which she discovered in the process, would yield benefits in terms of income and estate tax advantages. When a plan was elaborated Mrs Strauss and others of the group made an appeal for public support and received financing from the land-use committee of the local Farm Bureau. The Marin Agricultural Land Trust was incorporated and began operation in 1980.

Another example, although of smaller farmers, comes from Hancock County, Maine, where CLTs are '... working to restore access to land for local people who depend on land not only for

housing but as an immediate source of food, fuel, and employment opportunity' (White and Matthei, 1987:53).

Once a Community Development Corporation is established it
facilitates the formulation of other organizations to attend to
specific problems. Another of these that focuses on community
labor resources is the worker owned firm.

Employee ownership

One of the traditional attractions of urban areas has been
salaried employment. The 'company towns' of many countries
formed the base of community and family security ... until
recent years. With the advent of the 'post industrial era'
companies have consolidated, moved, or gone out of business
altogether causing great social and economic problems, especially in one-company towns. An answer to this dilemma has
been for a group of workers to organize and purchase the ailing
firm. Charles Turner (1987:77–8), a consultant for a firm that
promotes worker ownership projects, has elaborated a series of
requirements for success:

- **Leadership**. There must be at least one person in the
  community who knows the resource situation, the people and
  their interests, and understands the potential of worker
  ownership as an alternative.
- **Resource base**. There must be something to work with. The
  existing resources can be mobilized to secure those that are
  lacking. A responsibility of the leadership will be to identify
  needs – capital, technical assistance, communication technology, or political support – to coordinate procurement
  efforts.
- **Market**. There must be a demand for the product. Some
  applied research should be conducted to make sure that there
  is an available clientele to be served now and in the foreseeable future. It may be necessary to make product adjustments
  to assure satisfactory market performance.
- **Infrastructure**. A sound legal structure is critical for effective
  action besides serving as a guarantee to maintain the
  confidence of community participants. A systematic
  accounting system is also vital for all to be aware of exact
  details of the economic situation, their duty to it and their
  part of the benefits.
- **Skill and determination**. Responsibility, for a salaried worker,
  is principally for his or her own performance. While there are

reports of Japanese industries where employees sense responsibility for 'their plant,' this is more descriptive of Japanese culture than of work habits. The rules of the capitalist system include a clear division of labor. For workers who have been socialized in capitalism to participate in an employee ownership project, a change of orientation will be necessary. Responsibility is no longer for the use of personal skills alone, but involves accountability to each other. It may not be possible to just walk away from the plant at the end of the day. There may be meetings. Negotiation, dedication and a generous dose of goodwill are required.

- **Management**. Turner explains that managers need two critical skills: they must be able to build an organization that can compete efficiently in the market place and they need to be able to stimulate workers to assume responsibility for the development as well as the day-to-day activity of the firm.
- **Education**. In addition to the necessary recycling of skills that is especially important with accelerated rates of change, there is also general information about other aspects of the production process that are not usually encountered by a salaried worker. The 'holistic' approach often cited in recent times provides every employee with insight into the workings of the whole firm and furnishes a forum for individual suggestions.

Employee ownership of a firm, in order to provide supplementary income and occupation, is a useful concept for agriculturalists who have particular job requirements due to farm responsibilities that restrict their participation in off-farm employment. With owned firms there can be more sensitivity to other obligations and recognition of them in the formulation of work schedules. On the other hand, of course, farm chores would need to be arranged, as possible, to meet the outside responsibilities.

The third 'production factor,' in addition to land and labor, that is necessary, and often limited in rural development, is capital. There are community institutions for that too.

Community finance institutions

Rural communities around the world have been 'decapitalized' as resources moved in pursuit of urban opportunity. With the flight of capital from rural areas, the market becomes distorted and costs of basic goods and services increase. The result has been that community development, even when feasible in other terms, has been difficult to launch. An answer to this predicament is the establishment of local financial institutions. Michael Swack, President of the New Hampshire Loan Fund, has extensive experience on the subject. He comments (1987:80):

> *Community investment seeks not only to make capital available but to encourage the types of efforts and institutions that will make the best use of capital in the community. A successful program of community investment will stimulate the demand for development capital by supporting the formation of community groups and encouraging existing groups to plan more projects.*

Traditional money lenders may not identify with the types of need that exist in rural areas so as to be able to provide a fair and objective appraisal of loan applications. The result is that these areas are left behind. In some cases farm loans or rural community loans need deferred payments for several years until the investment begins to yield returns. There may not be the traditional equity against which to borrow. It may be necessary to visit the site and collect information to make a loan decision. These are all barriers to the cooperation of traditional lending institutions. But there may be sources of capital which are sensitive to the need in rural areas and would lend a hand if they could find a way. The two forms of organization commonly used to resolve this quandary are regional voluntary organizations and groups established within existing institutions or among individuals that have some form of 'common bond.' As examples of these two forms we will now examine the US models: Community Loan Funds (CLFs) and Community Development Credit Unions (CDCUs):

Community Loan Funds

The CLFs are non-profit corporations or are located within such corporations, frequently in Community Development Corporations. As local organizations they have specialized knowledge of the community, its resources and its needs. But CLFs are not simply groups in search of capital nor creations of investors in search of local investments. They include representatives of each of these groups in addition to technicians for planning and feasibility studies, and community leaders. The Boston Community Loan Fund, for example, has representatives of the CDC with which it is associated, bankers, representatives of religious organizations and community activists. The function of the CLF is to move capital, to educate the population in terms of local capital needs and opportunities and to influence economic growth and development.

Swack (1987:84) demonstrates the linkages among many of the concepts that we have discussed:

> By helping to capitalize community land trusts, limited-equity housing co-ops, worker-owned businesses, community development corporations and other community-based economic institutions, the loan fund alters established patterns of capital allocation that have made it both difficult to develop these institutions and difficult for those that do exist to prove themselves.

Community Development Credit Unions

The credit union as a source of capital is another community agency that functions within an established institution. In the case of community credit unions member linkage is based on residential bonds. Swack (1987:91) reports that 400 of the US's 19,000 credit unions are CDCUs. All credit unions in the US are regulated and supervised by the Federal Government which also insures deposits against loss.

The CDCU serves as a savings institution and lends capital for community needs. While some unions accept deposits from outside the organizing group, this is not the principal source of capital. A slate of officers is elected from the membership to handle routine decision making, a board of directors is responsible for policies, procedures and management, and a loan committee determines how resources will be invested. The organization is overseen by a supervisory committee that reports back to the membership.

Both CLFs and CDCUs have national organizations which unite the local units and can, if well integrated, move capital among the various member organizations for most efficient application.

The organizations that have been reviewed are examples of many others that exist all over the world and models for still others that could exist to alleviate problems in rural areas. New ideas and technologies increase the potential success of community action even where projects have been attempted in years past with limited success. Yet there is need for innovative ways of communicating new ideas to the community, the farm and for individual use.

SPREADING THE WORD

We began the chapter with a presentation of the concept of computer networking as a method of increasing communication among rural people. There are many other possibilities that can be considered. In various countries of western Europe local television programming provides a community forum for the discussion of local problems, school presentations, labor disputes and other topics of interest. At regional and even national levels, television is being used to make people aware of new ideas in agriculture. The program *Globo Rural* on national television in Brazil exemplifies possibilities for diffusing innovative ideas concerning methods for the increase of agricultural production and productivity, recipes for home preparation of agricultural products, interviews with farmers who have experiences to share and responses to viewers' questions. One series of program segments followed a cattle drive over several hundred kilometers from range to market. It picked up the human interest aspects of the cowboys, owners of land along the way and the hardships for the cattle ... and the television crew. Although this form of transport is no longer used, the series increased understanding of historical agricultural practices. It is interesting to note that the program, broadcast on Sunday mornings, is very popular and many of the viewers are urban residents.

Agricultural Extension Services around the world have waxed through well-planned professional programs that provided service unequaled by the contributions of any other rural agency; and with changing priorities they have waned as a result of budget cutbacks, consolidations and even, in some cases, cancellation. This comes just at a time when there is real need for the diffusion of communication technology and other innovations, some of which involve agricultural production and others that involve the people and their organization. How could the extension services, established in nearly all countries, be integrated into some of these new ideas? To be sure extensionists have discovered many ways to increase the dynamism of diffusing modern agricultural methods and practices in spite of reduced support for their programs. And there are other possibilities.

In some places it may be possible to hire local producers to promote specific innovations with which they have experience. With some assistance in the presentation of ideas, these temporary 'lay extensionists' would gain a new perspective on the work of extension. In Togo, West Africa, we invited village chiefs to send youth representatives to a week long training program on rabbit raising. Each youngster learned to use tools, built a cage, received instruction in rabbit feeding and care, and some techniques on how to teach new ideas to others. They went home with a female rabbit each and a male to be shared among those of each village. It was a project that produced great interest in the extension programs of the rural development center, as well as thousands of rabbits!

Another possibility may be the Agricultural Camp. During off seasons and once the corn and beans are laid by, many farmers have some free time for recreation. Religious groups have camps, youth groups use them to advantage and business men call them retreats. Why not farmers? There's certainly no lack of subject matter. The technical, social, economic and even political areas offer topics that could generate interest for several years at camp.

This brings us to some final suggestions. We need more communication: more talking and more listening. Farmers need to study the topics concerning our profession, formulate opinions and express them. At the same time that there are articulate agricultural leaders in all countries, many others have little to say. Recently at a Soybean Field Day a group of producers gathered around the university professor who was explaining the benefits of a new variety. As they returned to the tour bus one fellow stayed behind. The professor later related that when the others were out of earshot, he asked, 'Professor, let me ask you a stupid question . . .' We worry too much about self presentation, and perhaps too little about the future of our profession. Hopefully, with new information systems we'll talk more, listen more, and be listened to. It's a question of survival.

12

Earth Husbandry, an International Soil Building Program for the Next Century

The numerous advantages of family organized production listed in Chapter 3 are threatened by international economic and financial developments as well as an accumulation of mechanical and chemical innovations that are more suitable for large operations, create pollution and disorganize the farm population. We are mechanically following a systemic spiral based on profit and growth that cannot possibly continue. There are casualties – not only the splintered, displaced families that have been forced off the land, but those who remain, forced by investment and debt to perpetuate this system, as well as the urban population which has less contact with the processes of agricultural production and less possibility of consuming natural products of purity and high quality.

The social organization of production here examined has been essentially untouched by directed change and has degenerated as it has been buffeted about by other interests. We have seen some suggestions to improve the situation. Communication technology offers possibilities to reduce isolation and promote farmer organization. There are other forms of technology that could be applied to strengthen family agriculture.

In this day of lasers and satellites, of computers and telecommunications there is innovation that has gone untapped in terms of reorganization of the structures of agricultural production. It is now possible for the dedicated, conservative producer to be remunerated by society for his efforts in building our national resources. Unfortunately conservation has become a derided term, especially among those associated with production and economic development. But they use the term as if it meant preservation in 'unchanged' form. Conservation is a process. It involves active attempts to build toward objectives as opposed to allowing the erosion of natural processes.

If electronic and communication innovations were applied in a monitored initiative for resource building it would be possible to develop programs of a type that could, for example, be called *Earth Husbandry*. This would be a project not unrelated to the EU requirements to set aside acreage according to the CAP or the American Soil Bank program, but with more dynamic and long-term characteristics. Rather than putting resources aside as we put money in the bank, there would be an active production of resources. A farm family could, for example, contract to produce a 2 per cent increase in organic matter, or higher levels of soil nutrients in a specified area, over a five year period or longer. This is not to say that the area would not be cultivated, but the trade-off between producing grain and building soil would be remunerated and thus worthy of consideration.

Research (cited in Edwards *et al*, 1990: 260) has shown that:

- organic matter can reduce pest and disease incidence by increasing species diversity in favor of natural enemies (Altieri, 1985; Edwards, 1988);
- organic matter can promote populations of fungi that control nematodes (Kerry, 1988);
- organic matter can absorb and inactivate pesticides (Edwards, 1966);
- organic matter can provide alternative food for marginal pests and decrease their severity (Edwards, 1979).

Production of organic matter thus appears to be an important component for an innovative program of earth husbandry.

There are other advantages. With higher levels of organic matter the soil is able to retain more water. Reducing runoff reduces erosion, leaching and the transport of organic and chemical materials that would pollute streams and rivers. More capacity to retain water means more available water for crops, pastures and all other vegetation. As the organic matter is formed from green material that is incorporated into the soil, the plants, through the photosynthetic process, capture carbon dioxide and hold it in a fixed state. By increasing the 'green' we would compensate somewhat for the losses due to deforestation, and reduce levels of free carbon dioxide (CO_2) which are being blamed for the global warming trends. These trends are being documented so thoroughly that they can no longer be ignored. Al Gore (1992:38) in his book cited earlier, comments:

... our annual production of CO_2 and other greenhouse gases

is already so large and is increasing so rapidly that simply stabilizing the amount already in the atmosphere would require significant changes in the technology we use and in the way we live our lives. I suspect that many of those who say that it is probably all right to run these risks – to make no change in our current pattern – are really saying that they simply do not want to think about the disruption that would accompany any serious effort to confront the problem. Our vulnerability to this form of procrastination is heightened where strategic threats to the environment are concerned because they seem so big as to defy our imagination.

If soils are overused they can actually become part of the problem rather than a solution. The scientific relationship between carbon dioxide and soil organic matter is supplied by agronomists, Lal and Stewart (1992:5–6):

Soil degradation and environmental pollution go hand-in-hand.... In addition to the effects of deforestation ... it is now widely believed that the world's soils play an important role in the global carbon budget. Soil misuse and over-exploitation, causing rapid depletion of soil organic matter, can lead to emission of greenhouse gases into the atmosphere.

The possibility of taking land out of grain production for some time would allow the planting of trees for pasture, shade or windbreaks and bring up nutrients that have leached deeply into the soil. The great number of leguminous trees that would be candidates for planting, also suggests possibilities of a symbiotic relationship of nitrogen exchange with pasture or crops planted at a later time. In areas where it is necessary to fallow land to build up water resources vegetation could be planted that would be less demanding of water, while keeping the land in the production of organic matter.

The advantages are not all in improved soil. Earth Husbandry would provide areas that could support higher levels of wildlife. This would mean a more balanced ecology and better hunting to maintain the appropriate levels of wildlife populations.

More important from the standpoint of those in agriculture is that Earth Husbandry would provide an alternative. Farmers could increase their control over the decision making process as they decided among crops, pasture and soil building as economically feasible ventures. The freedom that used to be so noticeable in the US in adult farmers' meetings of high school vocational agriculture programs, before farmers were locked into intensive production to pay huge debts, could be returned. In those times the rustic market structure made it possible for farmers to create new possibilities of production. A man in our community castrated Holstein calves and fed them out for beef – when his neighbors were killing bull calves at birth. He pastured them in his woodlot and in fields away from public view because of the embarrassment. But he felt that he was doing something of his own invention that would put him 'one up on the world.' Even if he didn't make much money at it, although he did, it was psychologically important that he was able to try something on his own initiative and carry it out.

Today, an Earth Husbandry program might be considered useful to many to increase the flexibility of farm work and be able to take a part time job in town. By producing organic matter, rather than grain, the critical periods of farm responsibility would be reduced. On the other hand, the opportunity to produce a crop that didn't depend on market prices would increase stability and facilitate family budgeting. The message that soil building passes on to the next generation is a positive one: It's not just that 'we believe in the future,' but also that 'this is something for you.' A return to the concept of stewardship strengthens our historical linkages that have been important to those in agriculture all over the world.

If farmers who produced soil resources were given recognition in the same manner as those who produce high crop yields or outstanding livestock, more status and therefore interest would come to be associated with soil building. A recent article in the *American Farmland* magazine told of a US family on a diversified farm of 350 acres that produced nearly all their own soil inputs in the forms of green manure and livestock manures and sold livestock and livestock products as much as possible, rather than grain. They were integrated in a cooperative system with other families and the cooperative was, itself, integrated with other cooperatives. Sounds like a pleasant life, doesn't it? That family was recognized by the article and it served to motivate others who have similar interests and to encourage still others to join this innovative orientation.

An Earth Husbandry program is just an example of what could be done with satellite monitoring and coded telephone communications to regional computerized data banks, in the process of a dynamic program of production and conservation.

Hungary's agricultural system, as explained earlier, contains a component of 'cooperative farms' where the joint owners apply their labor and produce grain and forage that are subsequently transferred to their small household farms for livestock production. The farmers express satisfaction with this organization and are reluctant to change. Some have begun commuting to urban areas for industrial employment, but still reside on their own farms. From these farmers we can learn the importance of stability and risk reduction. Earth Husbandry could help in this too, not as a substitute for other forms of production but as a complementary activity that could be chosen as a useful application for resources.

EFFICIENT USE OF THE FACTORS OF PRODUCTION

Systems have been elaborated, in the developed countries, for the handling of capital in much more efficient forms than are available for labor or land; the other 'factors of production.' Excess capital can be invested in dynamic short-term projects such as the stock market, in longer term real estate purchases, or essentially put aside for the future, in bonds. For farmers there is need of this type of flexibility for the application of our labor and our land. It may be said that capital is easier to manipulate, to transfer and to negotiate. This is because the organization has been created and implemented for such activity. And this clearly is not true in all countries. On the other hand, if a similar effort were made in the elaboration of alternatives for the use of land and labor, the farm population would have a more adequate situation in which to apply modern principles of management.

Innovation could also be used, and is beginning to be used, to improve efficiency through increased knowledge of the real needs of the plants for fertilizer and pesticides. Application of an optimal quantity at the time that it will be most effective can reduce total amounts of inputs and machinery use (Pretty and Howes, 1993:28):

> Research at the Scottish Agricultural College is showing that if farmers get the timing of applications of fungicide on cereals right, they can cut rates by 50–75 per cent and still maintain yields. Researchers recommend that farmers regularly examine crops and apply a quarter-mix when 75 per cent of plants are showing at least one active mildew spot.... But these low dose approaches do place extra management demands on farmers. As Stuart Wale of the College put it: 'the use of low dose mixtures is not appropriate for all

growers. It is primarily intended for those who can inspect crops regularly and make a timely application of fungicide.'

'Smart' irrigation systems such as are in use in Israel have demonstrated a minimizing of inputs for an optimizing of output. A problem of this research orientation when it is applied to labor, however, is that the reduction of labor requirements for production hasn't made farmers' lives easier, it has eliminated farmers. The on-going production systems, that have themselves been a source of economic stability in the past have now become a threat. With Earth Husbandry, marginal lands could be brought back into some type of production using these modern techniques and retain more farmers on the land.

Keeping people on the land

Demographers study population movements using the theory of 'push and pull factors.' Farmers are pushed off the land to fulfill the objectives of large landowners to increase the cultivation of cash crops, or the objectives of governments that promote the 'grow or die' ideology associated with cash inputs and credit programs. This theory also holds that the population is 'pulled' from the land by the attraction of the urban life. The US song, 'How you goin' to keep 'em down on the farm, after they've seen Paris,' popular after World War 2, expressed the sentiment. And it was at that time that the flow of rural migrants increased to the urban areas of the US. But this too can be seen as the result of a 'push.' As we observed in the discussion of agricultural policy, families all over the world are determined to seek better health care and educational opportunities for themselves and their children. From an analysis of African agriculture, Barker and Chapman (1990:486) state, in a Nigerian publication, that 'migration to urban areas has been encouraged by the lack of investment in rural infrastructure and the subsidization of cheap food imports for urban consumers.' They cite a forecast: 'It is estimated that by the year 2010 approximately half of Africa's population will reside in the cities' (IITA, 1988:11).

Migrants in most parts of the world enter urban areas as second class citizens. They soon discover their lack of knowledge of urban ways, less and poorer quality education and limited skills that can be applied to employment. If health care, transport, educational facilities and production alternatives like Earth Husbandry had been available in the rural areas many, probably most, would have avoided the attraction of the city. Lack of effective programs on the part of governments and international institutions has left rural areas debilitated in their opportunities for service delivery. The establishment of useful production alternatives that would make it possible for those of us with farm backgrounds to continue in the area, would increase the economic feasibility of community services and strengthen family agriculture.

Earth Husbandry could also reduce the direct competition that farmers face in the market. The association between supply and demand, so that '... when there's less corn, I make more money,' has been distorted in the US in recent years by large exports to foreign markets. These shipments are being reduced, however, as countries that have made large purchases in the past become more self-sufficient. Some policy mechanism will be needed to occupy productive land while, at the same time, reducing US production and fierce competition in the market.

Individual gains and social benefits

There will be criticism of Earth Husbandry type programs, claiming that city dwellers are tired of subsidizing agriculture. This is a problem of vision. A soil building program is an investment made by both urban society and farmers. Short-term programs are more suited to individual gains. But in the case of the Swedish reforestation program mentioned earlier, it's less difficult to recognize that the forest doesn't 'belong' to its planter, but to society. We have all benefited from soil resources in the past, either as producers or consumers, and this situation, although more complex in today's international food commerce, is still the case. Doubters, go to Alaska and order a meal.

Isn't it preferable to pay the farmer to produce, rather than not to produce? Between dairy subsidies, a continuing struggle, and Earth Husbandry, a responsibility with guarantees, most of us would acknowledge the advantages of the soil building program. We would, of course, have to learn new techniques, some of which would involve unlearning practices of product maximization. And if we didn't get the job done (no doubt there would be cases of this) the satellite pictures would register a deficit in the contracted increase of organic matter. The process could be monitored with annual readings transmitted to local Soil Conservation Offices where data for the participating farms in the region would be stored. It would be a real learning process to check periodically to verify the expected accumulation of organic matter. There would also be a need for research in agricultural colleges and experiment stations focused on improved methods of soil building and plant genetics, to develop species with characteristics appropriate for the task.

It is important to remember that the trade-off between annual crop return and longer term income does not imply a loss to the producer. On the contrary, the organic matter that is accumulated will become a productive asset for future cash crops, in our own cropping systems as well as in those of our successors. We are essentially rearranging present productive systems to make them more sustainable as well as more appropriate to our present needs.

From an economic point of view, if grain production were reduced even slightly, there would be an expected increase in price. While individual farm income from grain may be reduced, it should be more per acre. And participation in the soil building program would augment income. Machinery, used on a smaller area, would last longer, reducing a major producer expense. This doesn't mean that it would be possible to turn the farm over to Earth Husbandry. Participation would, no doubt, have to be limited in the interest of spreading the opportunity. Nor does it mean that large industrial farms could use the program as a protection against risk. The program could be regulated so that they would have the right to no more participation than a family farmer.

A Global Marshall Plan

In his book, Al Gore calls for the mobilization of international forces to launch a Global Marshall Plan. Referring to the highly successful post-World War 2 plan that assisted in the rebuilding of Europe, Gore points out that there was more involved than the physical reconstruction of the bombed-out countries. The plan also encouraged specific political (democratic) structures and certain (capitalistic) economic orientation. His suggested Global Plan would also have an ideological underpinning. It would encourage ecological responsibility and resource construction. Gore (1991:295) admits that he is proposing an enormous project:

> Human civilization is now so complex and diverse, so sprawling and massive, that it is difficult to see how we can respond in a coordinated, collective way to the global envir-

onmental crisis. But circumstances are forcing just such a response; if we cannot embrace the preservation of the earth as our new organizing principle, the very survival of our civilization will be in doubt.

He continues to note that even a very successful program, in one country or another, would have little effect on global conditions. And so it is with Earth Husbandry. Our priority is family agriculture around the world. A notable advantage of this type of program is that it is appropriate for countries no matter what their stage of development. It is at least as interesting to a producer in Burkina Faso, who feels the Sahara Desert sweeping down upon him, as it is to a British farmer who could free up some of his resources in the short term to assure a long-term future. Colleagues who have reviewed the plan suggest that it offers more to those who have the least. Farmers in the developed countries who have eroded soils as well as Haitian farmers who are left with rocky hillsides that were benevolent to the long gone trees but offer little to annual crops, could be viewed as prime candidates for participation. This is not to say that the program would not be interesting to those with a history of resource protection. They may require the least convincing of all. Each country would need to establish its priorities and rules for participation. While the advantages might be posed in terms of resource construction and market stability in some countries, it may be pure income in others. There are areas where no cropping can be undertaken with assurance of results.

The underlying objectives of Gore's Global Plan and of Earth Husbandry, which could be considered one technical component of the Global Plan, are the same – survival. Gore sees our prospects dimmed by the overuse of resources and the build-up of pollution. Earth Husbandry seeks the survival of family agriculture as the productive structure that is most appropriate to rescue the Earth, or at least the future of its agriculture, for posterity.

The nuts and bolts of Earth Husbandry

Intervention programs require resources. As is generally the case with public projects, the economic aspects of any proposal present a critical topic for discussion. If we begin from the premise that resource depletion needs to be corrected, some will immediately respond that not everyone or every country has been guilty of the depletion. Gore (1991:176) cites British Prime Minister John Major as condemning the US leadership on environmental problems: 'The United States accounts for 23 per cent [of global CO_2 emissions]. The world looks to them for decisive leadership on this issue as on others.'

South American nations, conscious of the sources of global pollution, are unimpressed by criticisms from the developed countries that they are destroying forest land. A solution, to avoid hypocrisy and build support and goodwill, must begin at home. A European or American program could include many exemplary components. But other nations may require alternatives for reduced budgets. Gore discusses a program that has had some success between Brazil and the US, involving 'debt-for-nature swaps.' By this program the Brazilian Government promises the lending country that it will protect designated areas of debilitating resources in exchange for the forgiveness of part of its debt. It should be pointed out, in the case of Brazil, that the great majority of the national debt was accumulated during the regime of a military dictatorship when the people had no say in government spending. It is now a democratically elected government that is burdened with the enormous and growing, due to interest, debt. The debt-for-nature concept may be an affordable alternative in countries with sizeable debts. It is fundamentally based on survival objectives similar to those of the Global Plan and of Earth Husbandry.

Another innovation of the Global Plan (1991:345) is the establishment of carbon dioxide 'emission credits,' together with the possibilities of a market for their purchase and sale on national or international bases. This would be an international program by which national limits, expressed in credits, would be set through negotiation using technical knowledge and political persuasion. With the establishment of the treaty, the nations that successfully reduced emissions would be able to sell their credits to others that required more time for the adjustment. A variation of this idea might be useful for Earth Husbandry type programs. If it were possible for industries and other polluters to purchase the emission credits from the national government, within specified regulations, an initial source of funding would be created. It would not be wise for developed country polluters to be able to purchase credits from poor countries since, after some exchange of capital, the situation would likely remain as it is at present.

RELATIONS AMONG NATIONS

On an international basis, programs like Earth Husbandry provide new justification for foreign assistance. Traditionally foreign aid has sought to increase the access of the developed countries to raw materials, guarantee friendly trade relations, dump excess products (often agricultural) and a series of other unworthy purposes. The result has been that the developed countries are getting richer and the rest of the world is getting poorer. There are some bright spots, at least at first glance, among these apparently one-sided relations. The innovations in health care have been shared fairly generously in most parts of the world. Now in the less developed areas, principally as a result of effective vaccines and antibiotics, infant mortality has been greatly reduced, and the average life expectancy has increased. The results are included in Gore's survey: '... the world is adding the equivalent of one China's worth of people every ten years, one Mexico's worth every year, one New York City's worth every month, and one Chattanooga's worth every single day' (Gore, 1991:308).

What does all this have to do with Earth Husbandry? There is abundant evidence that those who are starving are willing to take great risks to migrate to places with more possibility for survival. People will not just quietly starve to death. Perhaps it is time to reexamine the present programs of foreign aid and consider projects that would build local soils and other natural resources in the hope of stemming the flow of migrants, both into the cities of their parts of the world and to the developed countries. There would be possibilities of increased food production capacity for the ballooning future generations, resources for economic livelihood and a higher quality of life for people in their own local environments.

Initial steps, in many severely degraded areas, would include the planting of trees to stop winds, shade the soil and begin building organic matter. An example of possible success has been demonstrated by Kenya's 'Greenbelt Movement' (Gore, 1991:324):

> Most of the seven million trees planted by the women in ...
> [the] movement have survived because a planter receives the
> small compensation for each seedling planted only after it has
> been sufficiently nurtured and protected to have an excellent
> chance of surviving on its own. The movement is now offering
> instruction on self-sufficiency in agriculture, and devoting

*some of the space in its nurseries to the development of seed
stock for gardens and fields.*

In the developing countries there are many species of trees
available that serve all of these functions. The leguminous
Leucena, already mentioned, and others offer nitrogen fixing
capacity and can be regularly pruned for forage. This increases
the total livestock carrying capacity of the land over the long dry
seasons. Acacia and other fast growing species produce firewood. But all trees (especially highly palatable, leguminous
species) are difficult to protect on lands that are being intensely
used. Such areas could be fenced off for Earth Husbandry
projects and still yield returns. Trees are also being used to
advantage in the developed countries. Researchers (Pretty and
Howes, 1993:31) in England suggest that 'alleys of trees' be
planted in cereal fields:

> *Recent research is showing that two-metre wide strips of
> mixtures of hazel, wild cherry, ash, sycamore, and walnut
> planted 12 m apart provide a good habitat for natural
> enemies, so permitting pesticide use to be cut. The loss of
> cereal area to the trees is partially compensated by an
> increase in yields of some 8% because of better shelter and
> moisture conservation, and 20–50 years later by the timber
> harvest.*

Another advantage of Earth Husbandry and this type of
technically monitored resource building project, is that they raise
the consciousness of producers to the reality of what is
happening to our productive resources. We hear so commonly
that six or eight tons of topsoil wash or blow off of each acre,
every year, in many areas. It becomes 'old hat' information and
isn't really internalized in terms of present and future returns.
The innovative technology that is now available, if applied to
agriculture, would make it possible to assess our conditions with
greater accuracy and would provide the data to convince others
of the seriousness of the situation.

Best of all, Earth Husbandry gives farming back to the farmer.
When we have resources deposited in a bank or investment firm
we expect periodic reports as to exactly how assets are doing,
what else is available, how fast, at what prices, values ... and we
may be upset if the report is delayed in the mail. A farmer's
resources are deposited in the soil. He too requires practical
alternatives, knowledge of how to mobilize them and the promise
of stability in at least a part of his enterprise. These are benefits

that could be offered on a global basis by a well planned and effective program of Earth Husbandry.

13

The Future of Family Agriculture in the Developing Nations

We've reviewed a series of topics that have affected family agricultural structures around the world. It's time to consolidate these ideas to envision what may be faced by the next generations.

The dialectical orientation that we have attempted to follow has taught us the importance of history in order to grasp the reasoning behind our present situation and begin to develop some thoughts about the future.

EXPORT EXPERIMENTS

Some nations historically became specialized producers of consumer goods and were able to export them around the world. Swiss watches and other artifacts did very well, at least until technological innovation changed the conditions of production and the ensuing advantages in the marketplace. Agricultural exports of a nonessential nature too; spices, coffee and sugar for example, have occupied a large part of international trade. This process gave the impression that export activity can be successful with any product in the market. In the short run this has been found to be true. On a more long-term basis, however, the relations involved, both within and among nations, reveal that agricultural staples don't follow the same principles. This is not for economic reasons, which is why some experts wouldn't agree. The difference is that these agricultural products, the essential part of our daily diet, are the means for survival. To survive we need production, and we need to produce. It is a question of feeding and clothing a population and also of creating gainful employment through which families can raise, or earn, the means to provide for their needs. No country is likely to become substantially dependent on another for its cereal grains and other foods considered necessary, over an extended period of time. As long as there are economic advantages in trade it will, no doubt, continue to exist, even in agricultural products. But often these advantages are apparent only from a limited, economic point of view. If real costs of social and ecological damages are considered, the apparent advantages are likely to be substantially reduced.

Even so, there is tremendous pressure from the exporting countries that are producing beyond the needs of their systems, for free trade, foreign assistance, and humanitarian efforts based on ideological discourse and economic enticements. Developing countries are told that there is no need to use their limited resources for basic research to 'reinvent the wheel.' It is usually suggested that these resources be invested in adapting imported ideas to local conditions. This doesn't solve the problems of royalty payments, however, and leaves the jobs in the exporting countries. It is also notable that the country that imports solutions before it discovers the problems, does not evolve the knowledge base that will permit future innovation. The possibility of discovering local solutions for local problems is thus reduced.

Ann Tutwiler and Barbara Elliott have written that even in 1980, 39 percent of US cropland was planted for export (1988:49). This situation too, has historical roots, which they well explain: after World War 2 the United States emerged practically unopposed in the international economy. Market competitiveness of US products was chiefly determined by their cost efficiency rather than variations in international monetary exchange rates. The dollar functioned as the worldwide standard. Under these conditions economic growth was practically guaranteed and US agriculture prospered.

During the energy crisis of the early 1970s several factors came together to increase the interdependence between developed and developing countries. The international flow of capital greatly increased, both in volume and in speed. With this came recognition that the old fixed rates of exchange between currencies no longer reflected the true value of each nation's monetary unit. There was a necessity for variable exchange rates among currencies. This is to say that trade in currencies – US dollars for Japanese yen, Brazilian cruzados for Swiss francs, and so on – would follow the rules of supply and demand as needed for economic exchange of goods and services, on a daily basis. Since the establishment of this system, exchange rates have varied widely at times, an occurrence that is explained in differing ways by financial experts.

In any case, agricultural production is valued in terms of the currency of the nation in which it is produced or owned. This means that when the dollar became more expensive relative to other currencies, the countries importing products from the US had to pay more. Given the unique characteristics of agricultural products it was not possible to simply cancel food imports, on which countries had accumulated a significant dependence, from one year to the next.

All this happened just at the time (1973 and following) that great quantities of 'petrodollars' were entering the international market. The petroleum producing countries were not able to absorb their increased wealth from the higher oil prices, and began depositing it in multinational banks. The result was an increased flow of capital that led to a growth in inflation and advantageous conditions for borrowing money. The developing countries were eager to jump on the merry-go-round of 'progress' that they observed in other nations. Tutwiler and Elliott (1988:56) observed that

> With available credit and extremely favorable terms, borrowing by developing countries rose dramatically. Total external LDC [less developed country] debt rose from approximately $70 billion in 1970 to $666 billion by 1981.

The spending spree that resulted was an economic boost to the economies, and especially agriculture, of the developed countries. They go on to say:

> The United States was a major beneficiary of this borrowing, as it fueled an increase in the consumption of all goods, but especially food. In 1972, agricultural exports to Latin America, Asia, and Africa accounted for just over 10 percent of all U.S. agricultural exports. By 1981, those countries accounted for 43 per cent of all US agricultural exports.

The developed countries then reacted severely to bring inflation under control. Interest rates began to rise, the value of the dollar increased in relation to other currencies, and the debts of the developing countries spiraled out of control. All this occurred together with reduced market opportunity for developing countries due to the global recession that followed. Attempts to augment local agricultural production were also frustrated by the fact that the modernization of agriculture required the use of imported goods – chemicals, machinery and equipment. The large debts and precarious financial circumstances of most developing countries led banks in developed countries to reduce loans. According to Tutwiler and Elliott (1988:58), 'Net bank lending to developing countries had fallen from $35.3 billion in 1983 to a negative $7.1 billion in the first half of 1986 (meaning the LDCs are experiencing a net outflow of funds).'

The reduced lending meant less money in the developing countries to spend on their most critical import: food. Reduced international food purchases meant economic problems for food producers in the developed countries. Families in developed countries were stuck with heavy debt, sometimes more than the total worth of their operations, and families in the developing countries had been driven nearly to bankruptcy by the cheap food imports.

The problems have not been exclusively between the well known developed and the developing nations. Even as some of the developing countries have built competitive industrial sectors they have caused problems for their neighbors. In Bolivia, for example, one can find black market goods from Chile, Argentina and Brazil for every need, at prices that would be impossible for the Bolivians to match. They thus become a repository for second quality, out of date or stolen merchandise from the neighboring countries. This climate is not encouraging for the production of any goods other than drugs and a few other agricultural crops.

WEALTH DISTRIBUTION: ITS IMPACT ON THE MARKET

When US President Nixon was making overtures to China there was talk of a billion potential consumers for American products. Today free trade advocates speak of 100 million potential consumers in Mexico. What is misleading about such statistics is that millions of poor people have very little to do with profitable economic ventures. It's not the numbers that promote trade, its what's happening to all of those people. In countries where there is a rooted aristocracy that controls the great majority of national resources it makes little difference, for economic possibilities, how many millions there are. They will probably remain at the level of subsistence, or below.

A group of researchers has recently provided a good example in studies that linked the marginalization of growing numbers of poor people to problems in the environment. They predict a future of violence (Homer-Dixon *et al*, *Scientific American*, February 1993:38):

> *The evidence that they gathered points to a disturbing conclusion: scarcities of renewable resources are already contributing to violent conflicts in many parts of the developing world. These conflicts may foreshadow a surge of similar violence in coming decades, particularly in poor countries where shortages of water, forests and, especially, fertile land, coupled with rapidly expanding populations, already cause great hardship.*

The problems thus develop due to an interaction of social forces with the environment. As pressure for fertile agricultural land increases, the ruling elite may change laws to favor themselves. An example is given by the authors of the Senegal River Valley in Africa, where most of the agricultural land is located that produces food for both Senegal and Mauritania (on the two sides of the river). The plains have traditionally been cultivated extensively as they flood every year, leaving a layer of rich alluvial soil that fertilizes crops and natural pasture.

> *During the 1970s ... the prospect of chronic food shortages and a serious drought encouraged the region's governments to seek international financing for the Manantali Dam ... and for the Diama salt-intrusion barrage near the mouth of the Senegal River between Senegal and Mauritania ... anticipation of the new dams raised land values along the river in areas where high-intensity agriculture was to become feasible. The elite in Mauritania, which consists primarily of white Moors, then rewrote legislation governing land ownership, effectively abrogating the rights of black Africans to continue farming, herding and fishing along the Mauritanian riverbank.*

Violence followed, in which black Africans were accused of violating the laws and driven out of the area. The lesson we can learn from this is to be aware of large agricultural projects functioning as possible mechanisms by which powerful people monopolize resources and poor people lose their access to land, resulting in what has been referred to as *de-development*. The dispossessed families move to the outskirts of urban areas where they contribute to the bloated urbanization, or they move to areas of the poorest soils, often vulnerable to ecological problems because of slopes. There they plant up and down the hills. The topsoil washes down the river and they continue to be poor.

THE QUESTIONABLE VALUE OF IMPORTS

Governments in developing countries have enthusiastically imported agricultural innovations for several reasons besides the allusive economic advantages cited above. Some have been based more on urban objectives to reduce food costs than on rural potential for development. Pressure is applied on farmers to reduce costs or suffer political, social and economic consequences. These threats come from their own political leaders: if they do not accept national conditions (prices, subsidies, plans for modernization) cheap food will be imported to further reduce local agricultural prices. The instability of production caused by this conflict leads to confusion and weakens the nation at a critical point of subsistence. The possibility of cheap food imports means that the 'dynamic' industrial, urban sector would be less dependent on the agricultural production of the rest of the country.

The growth in populations of developing nations has also stimulated forward-looking political leaders to encourage the importation of modern technology to increase production and productivity. But these innovations, as we found earlier, are not always adequate in the changed circumstances. Hartmut Schneider reports research (1984:103–4) which evaluated the results of tractor importation in Sub-Saharan Africa:

> The use of tractors may be important in terms of timeliness of cultivation but it is also the most discussed instance of employment destruction by machinery with, in many cases, a low or even negative social rate of return over cost ... problems of organisation, maintenance and inefficient use make tractorisation not only socially undesirable but also economically wasteful and inferior to cultivation by animal traction.

In most developing countries more jobs mean more access to food, a critical element for population well-being and political stability.

As exotic hybrid crops have taken the place of lower producing, traditional cultures the dependence on great quantities of chemical fertilizer and water has added risk to production that was already hazardous. The traditional crops, on the other hand, although less productive and sometimes more labor intensive, are more able to withstand climatic stress and can produce without high levels of inputs.

Most important are the personal conditions of producers' families. Imported technology assumes certain conditions that may not reflect the reality of developing countries. While some foreign technicians who have made brief visits have complained of producer inactivity – sometimes even speaking of laziness – they are generally unaware of debilitating health problems such as internal parasites from the water, that act up when people increase physical activity. Farmers may not be able to respond to the most technically correct recommendations.

An erratic supply of agricultural credit (time of availability, quantity, conditions) would stymie the most well planned program. It is not unknown, in some countries, for credit to be offered by the government only after planting time.

In places where groups of producers work together there are community activities (funerals, storm damage repairs, religious festivals) that take them away from their productive labors, sometimes at critical periods of cultivation.

These are all reasons why every nation needs to be involved in its own agricultural research at the same time that it shares information with others, preferably those experiencing similar conditions and circumstances.

While imported ideas may be well meaning they frequently cannot offer appropriate solutions due to their inability to reflect awareness and knowledge of local agriculture. Even the humanitarian organizations must be held in question. A recent statement made in the *Columbus Dispatch* (16 December, 1992) by Whitney MacMillan, Chairman of the CARE Foundation, expressed his advice for 'the long term solution to problems in Somalia and countries like it':

> Countries should produce what they produce best and trade. Comparative advantage is a fundamental economic principle. Subsistence agriculture, on the other hand, cuts off opportunities for economic growth. It segregates countries from participating in global trading.
> Liberalization of world trade will make it easier for all countries to achieve the economic gains that come from trade.

It should be noted that Mr MacMillan is also the chief executive officer of Cargill Inc.: '...a global, privately owned agricultural company.'

THE WEAKENING OF INDIGENOUS STRUCTURES

We have seen that imported ideas and technology are of questionable value in the developing countries. What about the structure of agriculture itself, the organization that each country chooses to produce the needed food and fiber? Our discussion in Chapter 8 examined plantations or 'estates' as structures that were implanted by colonial powers to produce for their own special needs and import the products. While this system was foreign to all of the. nations and territories in which it was established, it had some effect in bringing those areas into contact with the rest of the world. The flow of innovations – nearly all from the developed to the developing areas – resulted in increased production and, to some extent, progress. But enormous quantities of raw materials left the developing countries for processing, and furnishing jobs in the richer countries. Meanwhile the innovations from the rich countries that were accepted in developing areas created a dependence on their part for a continued flow of parts, technicians and complementary technology. The indigenous production structure, generally some form of family centered organization, was not strengthened by these activities and, in fact, faced stiffer competition in the market. As plantations have been phased out the family structures have had difficulty taking over production on a commercial basis. Urban populations have grown so rapidly that without imports there would be food deficits and consequent social and political upheaval. The critical question becomes how agriculture can be changed to meet the national needs. But it is more likely the nations that must change, in terms of the proportions of their people who live in cities.

In some nations drastic steps have been taken to reestablish family agriculture. Christian Andersson (1985) studied the agrarian reform of Algeria in the 70s. He explains that more than one million hectares were distributed to 100,000 beneficiary families (approximately 10 hectares or 24 acres to each). But the program sought to meet its objectives by force. Cooperatives were established by the government and membership was compulsory. Compliance with program objectives put families in a position of heavy dependence on the government. Many lost their initiative as they simply carried out the mandated program. As a result, Andersson reports, Algeria produced less cereal grain in the 80s, with a population of 20 million, than it had at the beginning of the century with only 5 million people. National food production, which had met 70 per cent of consumption needs in 1969 (before the agrarian reform), met only 30 per cent in 1983. The deficit was purchased with returns from petroleum exports. But the author questions whether petroleum offers the future security that would be probable from a strong agricultural sector based on family agriculture.

SURVIVING 'AS A PEOPLE'

Nations with large foreign debts involving financing from the World Bank are subject to monitoring procedures based on standardized efficiency criteria which force them into a western mould, not necessarily sensitive to national values, principles or objectives. The allocation of government funds is reviewed with strong recommendations as to where investments should be made and what steps need to be taken for debt repayment.

A common myth in the developed countries is the idea that the western fast food chains, clothing firms and various other 'symbols of development' are welcomed in the developing nations and will, with time, take over local markets. To be sure there are those who enjoy imitating westerners, but the results of attempts at world standardization are not likely to be positive for the local processes that are essential to socioeconomic and cultural survival. This is true even in the developed nations.

A recent meeting of leaders of the North American Cherokee Nation was held in their former homeland, near Calhoun, Georgia. Deputy chief John Ketcher explained on US National Public Radio that '... we came here to celebrate the great spirit and determination of the Cherokees to survive as a people.' What does it mean to survive 'as a people?' Is it only ethnic minorities who worry, or need to worry about surviving? Is there any relation between the ethnic and geographical ties of families and forms of agricultural production? The answer is yes. A current example is unfolding in the Balkans

The Slavic republics were formed after World War 1 by politically joining various small kingdoms in the region. The people, however, maintained their separate ethnic and religious identity which included, until recent disturbances, family garden plots and vineyards for wine making, even among families living in urban areas. As Frank Klicar reported in his film *The Yugoslav Republics* (1991), the justification for family food production is generally given as being economic and because of differences of taste. But the processes of production, storage and preparation have helped maintain their ethnic identity 'as a people' over the years. Meals taken outside the home are considered 'experiences,' observations and evaluations are made for later comment.

It is in this atmosphere that, when political oppression was relaxed, the various groups that made up the 'nation' returned to their ethnic roots and sought separation from one another. It may be argued that the national unity was superior to the great suffering and disturbances that have followed. But the ethnic identity of the people is an antithesis that exists (whether 'good' or 'bad') within the national structure and has preserved identity as well as preserving the will of the people to continue their lives in an ordered, stable manner. The individualism spoken of by the French in Chapter 2 'frees' us from what may be considered unbearable or even unpleasant situations. And individual freedom also 'liberates' us from our cultural heritage. This could be extended to 'freedom' from the family itself, and ultimately from any future that would be recognizable to us who live in the present.

Ethnic identity and the nation state

The political procedures of carving out nation states, as was done in Yugoslavia and all over the world, have joined peoples of different traditions, customs and religious persuasions. As long as times are good economically, methods are found for peaceful coexistence among different groups, and if the government is strong and acceptable, the nation survives. In West Africa nations were sliced out of the lands which belonged to them by the European colonial powers in North-South wedges below the Sahara Desert. But the local people had historically migrated around the southern edge of the Desert along East-West lines. The result is that the same nations are found extending across several countries. Some have, as determined by European rulers, English as their national language. Others of the same ethnic group but residing in a neighboring country, speak French. But when asked their nationality, most adults will respond with pride, citing their ethnic origins, not their political residences.

Agricultural methods, food crops and food habits have been important to these people in their struggle to retain their identity and differentiate themselves from their neighbors. The climatic similarities have also helped in keeping them under similar conditions. Since climatic variations also exist in East-West zones, ethnic groups generally produce similar foods among themselves, and different than that from other groups. Even staple foods vary, for example, from sorghum and millet in the dryer areas closer to the desert, to rice in the coastal regions further south.

Farm ethos and social continuity

Ancient burial grounds in various parts of the world have been found to contain animals placed together with their owners as well as commonly used tools and other artifacts. These symbolic practices reflect the underlying sentiment of the societies at the time. The fundamental character of past societies has been intimately associated with families and with agriculture. The socioeconomic stability that resulted when families were able to work together, producing for their needs and trading surpluses to secure what could not be produced, has contributed significantly to the perpetuation of humanity.

In times of social upheaval governments have used family agriculture to stabilize national conditions. Kenya was cited in Chapter 8 as an example. More recently some Asian countries have returned to family production for similar reasons. Daniel Kelliher (1985:307) writes that the coalition organized by Deng Xioping in China, after the death of Mao Zedong, 'allowed unsanctioned innovation to occur to encourage farmer experimentation and to stimulate agricultural production.' He admits that there was also concern that the great heterogeneous masses of rural people not be further antagonized by government action. Policy evolved that followed local initiative. A system of private land rental was implanted by the government. Families began producing on a subsistence basis and selling surpluses. With time rural credit became available and there was increased commercialization of farm products. The result, as reported in Chapter 2, was that food production has doubled on a per capita basis since 1970, in spite of a population of more than one billion. Hoping for similar, positive results, Mr Yeltsin has recently decreed that land can be owned privately, purchased and sold in Russia.

In India the rural development program that carries the name of the great leader, Mahatma Gandhi, stresses village self sufficiency first and village industry to absorb excess labor. The goal is to make it possible for families to remain at home, reducing rural-urban migration.

As developing country agriculture has followed the example of the plantation, specializing in a limited number of crops for exportation and importing for needs not immediately or as efficiently produced locally, we have strayed from the concept of local self sufficiency. In some places this orientation may continue with success. But regions that have become dependent upon imports for convenience will need to rethink their choices in the future. There is little probability that foodstuffs will continue to be donated in the same quantities as in the past. The exotic taste for wheat bread, that has developed in areas where wheat cannot be raised, due to cheap foreign wheat, is not sustainable.

The transformations that we have observed in societies are, of course, occurring in family agriculture too. Patterns of diversification using annual and perennial crops and livestock as well as part-time employment in the village or larger nearby urban area may be part of our future. Possibilities of 'value added' increments to agricultural products to increase farm income represent opportunities in many areas. The sale of meat rather than animals, cheese rather than milk and bread rather than grain are examples of using farm labor to increase returns.

A sad fact of our review of agriculture in the developing countries is that there is hunger in our future. The accompanying strife that brings violence and injustice is already appearing in the poorer countries. This is not a justification for invasion by the developed nations. All countries today have educated leaders who will plan the most effective solutions within local political parameters. To the extent that some groups are neglected by the politicians they must organize their own manifestations to secure justice. The appropriate role of developed nations will be to avoid the temptation to dump surplus goods, whether they are military, agricultural or any others. It will also be increasingly important to avoid creating dependence on imported staple products, that stimulate population growth on a false foundation and create enormous future suffering.

CONCLUSIONS

Our review of the historical antecedents in the developing countries demonstrates that the present situation has resulted from a specific set of circumstances that, in some cases, no longer exists. We can thus have little hope that the future will be a simple continuation of the past. We have arrived at a time in which the changes in production and in population are indicating an imminent qualitative transformation.

Recommendations from developed countries that all nations should specialize in what they produce best and trade to meet their needs is an elitist position that best serves the interests of those who have goods to trade, and the traders. History is full of resourceful nations' alliances to avoid facing difficult national problems. Even within nations, the residents of geographical areas have sought to separate from their poor compatriots – Kinshasa, from the Congo; São Paulo, from Brazil. Industrial sectors have complained about 'traditional' agriculture as a burden to modernization at the same time that they organize to increase sales of urban produced technology to farmers. But the disadvantaged segments of each country's population will not go away, even if they can be racially segregated, as was the attempt in South Africa. They may be burdensome, if ignored or exploited, or they can provide opportunities for the nation to build its productive capacity as well as to strengthen their own potential to participate in the consumption of goods and services. These are national concerns that need to be faced within each country. International commerce can undermine national progress and needs to be approached with care, understanding and based upon well established and accepted plans and national priorities. An essential component among these priorities is agricultural production. The promotion of family agriculture as a feasible means for producing both food and useful activity is a part of the solution in the developing nations.

It is not proposed that attempts be made to return urban residents to rural areas. An emphasis on retaining the present farm population on the land is a more attainable objective. As a sector that can more effectively accommodate dependent groups within the population, such as the young and the aged, there may be some return to farms as they provide better living conditions than urban areas, and as rural communities provide the safety, education and health services that are required. In the case that these steps are not taken, the opposite forces come into existence, and families encourage their children to seek a better future in the city.

With the end of the era of plantations in developing countries and the clearing of the last of the frontier lands, there is a tendency for farm size to be reduced. This is not to say smaller farms are to be recommended. Remember, our concept of family agriculture does not imply small farm agriculture. When there are advantages to economies of scale through mechanization or other innovations, they must be employed for the economic security of the nation. As there are possibilities for the industrialization of agricultural production this too must be investigated and put into practice, if determined to be advantageous to the population. The family structure should not suggest limitations for agricultural development. On the contrary, the strength of the family-based structure is the foundation upon which development can be constructed in the most diverse and appropriate manner that can be brought into practice by the populations of each nation.

Finally, the minimal levels of farm mechanization, a characteristic that has restricted social contact in developed nations, and the collective orientation that is common in developing nations, may actually furnish the environment in which cooperation can be encouraged on a more formal basis and intensive programs of resource protection implanted for the benefit of future generations.

14

The Future of Family Agriculture in the Developed Nations

Our discussion through the series of chapters and topics has, hopefully, provided information for forming opinions as to what the reader envisions to be the most appropriate activities for strong agricultural sectors in the developed countries. In this final summary we can attempt to apply the principles of dialectical logic to the subject matter that has been reviewed. As we have stressed the importance of relations among evolving entities we have entered into discussion that may have seemed extraneous to family agriculture. Communication technology, satellite monitoring, and, perhaps most of all, a discussion of logic may have taxed the reader's patience. The objectives: to describe what family agriculture is on a world basis, to determine its importance, and to suggest means by which it can be strengthened, mandated just such an innovative approach. The developed nations, where the family as we know it is most threatened, requires special examination.

We have seen that past prediction for farms to become larger and larger and more highly mechanized has been true only up to a certain point. There are suggestions that we have reached the possible economies of scale, at least for the foreseeable future. On the other hand the food industry has shown surprising growth as it incorporates more workers into salaried positions to provide for food needs. In general we could say that the 'food sector' as a whole is actually not increasing production with fewer workers. It has moved workers from farms to industrial centers so that the products are marketed in 'value added' forms, for higher prices. This is part of other processes occurring in these countries as there are increasing numbers of two-salary families with less time for meal preparation and more cooks who have never heard of broth and count the microwave oven as the essential kitchen appliance. With these developments the agricultural sector is more integrated into society. There are higher levels of standardization, and expanded control. The independent production of family agriculture is threatened by these processes and we might expect some form of meaningful, protective policy from governments. But our analysis of agricultural policy in most developed countries yielded some real questions: have government measures tried to respond to low farm income and prices, or have they simply sought to increase production?

Industrialized foods will, no doubt, continue to increase in importance. We saw an extreme example of this in the products based on fungi, in Chapter 7. It would be consistent with historical developments if food manufacturers, like those of other products, began producing their own raw materials in attempts to eliminate dependence on agriculturalists as much as possible. It is unlikely, however, that there will be reduced demand for traditional farm products. On the contrary, innovative products are being developed from traditional crops, as well as new crops introduced to meet modern demand.

In the face of this turbulent change it is especially important that programs of well defined national goals be formulated for food and agriculture in ways that are agreed upon by as many of those involved as possible. Only with the establishment of such goals will it be possible, for example, to determine priorities for agricultural research and the extension of recommended practices. Dahlberg (1986:8) cites a Rockefeller report (1982:9) on US agriculture:

> The lack of a coherent national agricultural policy, relating productivity goals and domestic and international policies with an explicit understanding of the value of agriculture to this country, greatly hampers efforts to establish national goals and priorities for agricultural research.

As is implied, goals should be based on knowledge of the situation, and also on national values related to agriculture. While knowledge provides understanding of the relative importance of profit, sustainability and need satisfaction in a just social order, still the quantitative analyses of statistics omit the values that provide us with meaning. They are, thus, only part of the information necessary for deciding what forms of production should be encouraged and how funding will be spent. Values, the significance of the family, survival 'as a people,' safe, pure foods, freedom, attitudes toward work, are all associated with the kind of countries that we are evolving into, further down the road.

LEARNING FROM THE PAST

History tells us that agriculture is a source of stability in society. It tempers economic events, reproduces the products, the people and their ideals and pacifies the social atmosphere. At the same time that agricultural surpluses have provided the resources to construct urban industrial areas, the sector has the flexibility to reabsorb excess labor in times of crisis. Capital invested in farm land, in general, holds its value, even as other investments crumble. Still, the structure of agricultural production in the developed countries is changing. The urban objectives of 'get big or get out' are encouraged and many of us have questioned the permanence of traditional family structures. But there are two questions from our discussion that still require answers. If we accept the growth model, with its emphasis on more with less, what is to be done with the increased production? And secondly, what of those expelled from the land? We have seen that international exports, on a cash basis, are likely to have problems, even in the near future. Food production is on the increase in all countries with agricultural and financial resources. A few nations remain interested in trading natural resources (for example, petroleum) for imported agricultural products, but they are not regions with sizeable populations. While there is hunger in many areas, those countries have limited resources. Imported food only delays and adds to impending crisis.

Although the developed nations have been successful in the past in reducing the size of farm populations, the possibilities of creating more jobs in urban areas are no longer hopeful. This is especially the case in the countries that are reabsorbing military personnel and those employed in military projects into the civilian population. It doesn't help to respond that 'this always happens when there are changes in the organization of production,' or that 'these are temporary disturbances, until the workers are absorbed into other sectors of society.' Most industries, like agriculture, are attempting to increase productivity and reduce the number of workers. We can only conclude that *growth is not a permanent opportunity*.

In the face of reduced future growth we must develop an appreciation and recognition of what we have. There is need for some gleaning of our present traditions, structures, methods and customs to determine our best options for a sustainable future with the possibility of hopeful life situations for the generations that will succeed us. And among these decisions we must consider the future of family agriculture. If we decide, as societies, that the family is a useful structure to produce for our needs, that decision must be accepted in all sectors and reflected in agricultural policy to reconstruct the status of the farmer and the aura of farm life. We have seen that the policies formulated by governments have great influence on farming structure and operations. People's lives are affected. Educational programming, mass media attention and parental concern can unite in a movement to assure those in agriculture that there will be economic equality and political justice for those who till the land. Among farmers there must be an understanding that the present problems can be solved when society puts itself to the task and that agriculture does remain a sound career possibility.

Together with the recognition of the importance of family agriculture must be guarantees in the form of agricultural policy, as mentioned above, with programs for adequate research and extension of the evolving innovations that will keep agriculture in rhythm with changes in other sectors of society.

Martin Kenney and his colleagues (1989:131) have examined the history of US agriculture and its transformation. With the assistance of new forms of theory elaborated in Europe they describe, in very convincing terms, the factors that have determined the present situation. Among their findings:

> First, the motion of the non-agricultural economy largely determines the shape and structure of agriculture. Second, the current crisis in agriculture parallels the current crisis of American capitalism and the origins of both crises are to be found in the institutions built during the New Deal and immediate post-war era. Finally, the linkages forged between non-agricultural industries and agriculture have resulted in a present day political economy in which the two have become entirely intertwined and inseparable.

We read very little that relates the agricultural situation to national economies and world processes. Are we ignoring information that exists on the topic or is there a lack of focused material in this area? Linda Lobao (1990:32) has concluded that '... much of the contemporary literature is built upon a narrow theoretical agenda focused on the persistence of family farming rather than its transformation and on the internal dynamics of the farm unit.' She, too, stresses the point that changes in all sectors of societies have implications for agriculture. It is thus not possible to delineate the optimal situation for farmers without considering the holistic socioeconomic changes of the world community.

Many of the changes noted by these authors identify modifications in agriculture as part of the conversion of farming from a life oriented structure to one of profit orientation. As such it enters the field of competition on this more narrowly defined basis and is vulnerable to being criticized as outmoded when more profitable structures evolve. As they recognize, this is a partial type of analysis that ignores aspects of sustainability of resource use, reproduction of competent personnel and other characteristics that we have noted.

The importance of efficiency as a goal is not being called into question here, but efficiency is not as simple as it may appear. The production of agricultural goods is an activity involving a combination of resources in systems of human-resource organization all over the world. It occupies the time and other resources of the people in these systems in ways that complement and compete to maintain the ongoing content and vigor of the systems themselves. The fact that one or another product can be produced more efficiently in another system at a given time isn't necessarily relevant to the total context of a system in question. By making decisions on the basis of isolated, economic criteria, distortions can be created that inhibit the development of the importer in terms of actually becoming a system that could produce equal to its inflated levels of consumption – increased by the imports. This is not to question the soundness, for example, of the Japanese exporting industrial goods and importing food, although they would probably not carry this to the point of significant dependence on imported foodstuffs. The reference, rather, is to countries like those of Africa, for example, that are increasing their populations (altering levels of system resources) on the basis of foodstuffs that are artificially increased through imports.

Another example is the development of computerized systems built of imported components, creating greater distance between the modern, electronics-using segment of the population, and the others. These distortions create stress and systems problems. The dilemma was addressed in our discussion of the developing countries (Chapter 13) but is reiterated here because of the importance that those in developed country agriculture understand the problems created by importation. As stated before, family producers, both in the exporting and the importing countries, end up being hurt by regular movements of agricultural products from one nation to another. The partial information disseminated in exporting countries yields great hopes for better times to come. These are transposed into higher land prices, and more difficulty for young farmers to get started. Producers in importing countries are eliminated from the market by prices that are unrelated to the costs of their operations. These are just two examples of the problems generated by the profit orientation. From the contrasting concept of a life oriented system, however, family agriculture can be understood not only as a means to socioeconomic objectives, but as an end in itself.

THE FAMILY AND THE FUTURE

Another factor that affects agriculture is the changing status of the family. Recently much has been said and written in the developed nations about the significance of the family to provide resistance to social maladies. The socialization of children within the family unit as well as the stable situation enjoyed by the parents provide a basis from which the tasks of production can be approached with objectivity. It would be simplistic (positivistic) to declare that a breakdown of the family is 'the cause' of social degradation. There are multiple causes as well as effects that themselves become causes in the extended time period. But strong families are one part of the solutions.

Clearly, the family that also works together and learns from one another grows with the assurance that, with careful attention to resources, there will be the possibility of carrying on in the professional tradition. Much is said in the developed nations today about mentors assisting young people into adulthood and the professions. This is in recognition of the importance of the human element in the processes of growing up. Centuries ago apprenticeship systems were highly formalized to train young people into professional responsibilities and pass accumulated skills from one generation to the next. Apprenticeship is still used in developing countries for these purposes. Currently, in the developed countries, mentoring programs are being established to partially substitute for absent fathers and guide young people through the decision making of maturing.

Family agriculture has a built-in mentoring system which is part of its strength. Usually parents, but often aunts, uncles and grandparents play significant roles as models for the children. Admittedly this was easier fifty years ago when it was common to sell butter, eggs and beef directly to consumers and families participated more completely in production processes. The activities of processing and distribution lent themselves to the participation of family members. Today most farmers in the developed countries have become producers of raw materials for processing in other sectors of society. Ironically this is what the developing countries complain about, recognizing that it is the processing that yields the jobs, the profits and the political power.

ORGANIZATION IS THE KEY

So how can farmers in developed countries avoid the pitfalls that have contributed to keeping producers in other parts of the world in the back seats of progress? There are numerous special interest groups working on these problems and well developed possibilities for communication. In addition to producers with interests in the protection of varying commodities – dairy farmers, soybean producers – there are those with an interest in reforms to promote, for example, the permanence of family structures. Other groups represent the processing and distribution of agricultural products through agribusiness, which may conflict with priorities of producers. Still other groups are interested in controlling what they perceive to be the negative consequences of food production and processing (Browne, 1988). But William Browne (1988:192–4) provides a word of caution:

> As policy decisions have become more specialized in terms of their agricultural purpose, groups have become less inclined to raise divisive philosophical questions about whether multiple issues and provisions are dealt with as a coherent policy whole. On the contrary, increasingly narrow groups have sought their own particular niches in hopes of being able to influence some small part of a total puzzle ... the impact of the exploding universe of agricultural interests is producing even greater incentives for even narrower policymaking, a common problem for democracies where the articulation of many interests overloads the policymaking process ... the multipurpose general farm organizations were found wanting because in an increasingly complex and technical agriculture they helped create partisan impasse.

To improve the organized protection of agriculture there is a

need for some form of linking among the various special interests. Workers in the industrial sectors are advanced in these
processes. Although employed in varied and sometimes competing industries, their union organizations give priority to
common interests, contribute heavily to political campaigns and
parties and constitute a solid block of political organization.

Perhaps a feasible option to unite the diverse interests of
agriculturists is some type of coordinating organization that
would take the initiative to communicate varied and divergent
interests among the factions of that sector. It may be that such
organization will be most successful if it originates with the
producers themselves. As farm people we have a rich history of
political action in benefit of our professions. Don Davis and
John Gaventa (1991:2) have studied these local activities in the
US and report: 'The political actions of ... emerging grassroots
groups, say researchers, demonstrate that ordinary citizens,
through organized collective action, are capable of influencing
public policy formation at both the state and local levels.' They
emphasize that as governments and private corporations have
failed to produce sustainable development on an international
basis the role of non-governmental and community based
organizations in shaping and implementing development
programs has become of critical importance: 'There is ...
evidence that this is an organizational possibility: in recent
GATT negotiations, local farmers from Kentucky joined
thousands of other farmers in Brussels to protest proposed
GATT policy' (Davis and Gaventa, 1991:30–1).

While organizational methods are more generally known and
used in the developed countries, the diversification of interests
has splintered efforts at unification. It may sound contradictory
but the goals of community development programs, at the same
time that they are conscious of, and in cooperation with,
processes at all levels, need to stress self-sufficiency. This is the
essence of the locally based, grassroots controlled organizations
recommended by Davis and Gaventa.

This does not imply that local programs should not look
beyond their communities to broaden their success as, in fact,
they have in some places. Pretty and Howes reported in Chapter
9 that in Australia a program that began with grassroots
organization is achieving great progress. The Landcare Program
involves more than 1400 communities working to develop
sustainable programs of land use with the support of the federal
government (Pretty and Howes, 1993:44):

Landcare comprises groups of farmers working together with government and the wider community to solve rural problems. It is an ethic of environmental responsibility, and embodies an 'education and persuasion' rather than 'legislation and coercion' approach to sustainable development. One State coordinator put it this way: 'Landcare is about getting groups of farmers together to tackle common problems. It is for the government to provide funding assistance and technical advice, but for farmers to make their own decisions.' *The Programme has achieved great success, but the factor noticed by commentators, local people and farmers alike is the sense of cohesion brought back into rural communities.*

Diffusion of information concerning related interests is needed not only in the seats of political decision making but at all levels of production-processing activities. We have reviewed some exciting possibilities for communication that are already in use in many parts of the developed countries. They may become of special significance in the unification of modern agricultural sectors. The research of the Floras, cited in Chapter 8, provided an example of the problem. Agricultural policy stimulated high levels of mechanization that, in turn, sent farmers looking for more land to increase the efficiency of their new investments. This process leads to higher land prices, debt, and eliminates families from the land.

UNITING THE RURAL COMMUNITY

We noted in Chapter 11 that the rehabilitation of rural communities also involves those non-farming residents who share the same geographical region. They too are needed to work for the inflow of commercial enterprises, capital resources and political representation. With increased activity there is flexibility for everyone to work to meet his and her personal needs in ways that are most compatible with the particular requirements of their family situations.

A word of caution: community development programs have sometimes been led by local businessmen who hope to turn personal resources into renewed forms of profit making activities. As such, the participation and multiplier effects of job creation and improved services for all rural community residents are limited.

A chief resource of most agricultural areas is labor. Industry will seek out possibilities to reduce labor costs, even where others of the 'factors of production' (land and capital) are more expensive. Companies that are currently moving to Mexico and other developing countries are not looking for cheap land and certainly not counting on foreign capital. They invest, using their own capital, in order to take advantage of cheap, unorganized labor. If they succeed in training local labor forces to improve efficiency and productivity and instil the dedication that is common among workers in the developed countries, and if workers make these transitions without demanding higher salaries, there may be reason for concern in the developed country workforce. Past experience, however, demonstrates that it is more likely that these industries will return to the developed countries. This is not before the industrialists have accomplished their goals – which include short term profit and the threat to developed country workers that alternative sources of labor are available when costs become 'unreasonably' high. By monitoring these processes we can plan for the future of rural communities in the developed countries.

With increased communication throughout the agricultural sector there are stimulating possibilities for organization and cooperation. We have seen that the motivation for group action often comes from specialized bonds among individuals and families: the Amish and Mennonite communities of various nations are examples. Native American populations are demonstrating notable ability to organize and defend their interests. We in agriculture need to put into practice techniques of modern community organization and development to strengthen our communications – and telecommunications – and increase our participation for the good of agriculture in general.

Hopeful signs are beginning to appear. Among the objectives for community development, health care is an example of great concern. As long as rural residents have access only to severely limited services and pay more for second quality care, there will be implicit encouragement to move on to larger, urban areas. In some regions just the possibility of health care of comparable quality and cost to that of larger cities would be enough for people to stay put. It is these 'high visibility' type problems that mobilize the most popular reactions. Waste incinerators that noticeably foul the air and water are another example that is garnering reactions, even among those with little ecological sensitivity. But we need to identify priorities and mobilize behind issues of importance, regardless of their visibility.

THE MOST IMPORTANT FARMERS: THE FUTURE FARMERS

Perhaps the most convincing argument for family agriculture is the product of greatest significance to us, our children. It is not surprising that children raised in conditions where they observe and participate in difficult tasks sometimes, and pleasant moments of leisure at others, are prepared for the uncertainties of adult life. The importance of overcoming obstacles and working in cooperation with others is now being simulated for youngsters in field trips, camps and classroom situations as it has been recognized to be an essential aspect in the development of character. Native Americans in the western part of the country have recently increased emphasis on the participation of their youths in work activities. A recent report on US National Public Radio explained that teenagers are being given active parts in rounding up the buffalo in the fall. Part of the traditional knowledge that is passed down from the elders speaks of the habits of buffalo, their intelligence and interaction with horses. In the round-up the youngsters get a chance to use this information. Adults happily report that they have noticed a reduction in juvenile delinquency since their children have become involved in this activity that has so much historical significance to them.

Another advantage of rural upbringing has been documented by a team of medical researchers in Europe. Glyn Lewis and his colleagues (1992) studied a group of nearly 50,000 Swedish draftees to determine the importance of environmental factors to mental health. They explain that previous studies have emphasized the genetic backgrounds of patients and accepted the 'geographical drift' hypothesis which holds that those with mental problems tend to drift into cities in search of help, resulting in a higher prevalence of such cases in urban areas. Examining specifically cases of schizophrenia, the researchers controlled the influence of other factors such as use of drugs, family finances and social ties and discovered that 'Incidence of schizophrenia was associated with place of upbringing. There was a strong linear trend, with the highest rate of schizophrenia in the cities, intermediate rates in large and small towns, and lowest rates in country areas' (Lewis *et al*, 1992:138). They concluded: 'Stressful life-events are associated with the onset of most psychiatric conditions including schizophrenia and are more common in cities.'

As the advantages of farm life are remembered in developed societies there is hope that family agriculture will be encouraged as an alternative career decision. When even a small proportion of a population pursues vocational choices that enhance certain aspects of the national values, these values live on to influence us all. To be sure, as farms become 'larger than family' operations the contribution to values is reduced. When cows are given numbers rather than names, when three hens are stuffed into a cage to become egg producing 'machines,' the life orientation comes under stress. Often it's the children who give names to animals, places and objects that stick in the family nomenclature. And just the fact that they can designate what objects are called gives them recognition as members of something that is important.

KEEPING UP WITH THE TIMES

Today there is great consciousness of the need to recycle resources and apply conservative practices. Families in agriculture, with their traditions rooted in history, have responded. Recycled newspapers are being used to bed calves; winter cover crops, like hairy vetch, fix nutrients to reduce the need for chemical fertilizers and control soil erosion. Rye grass is being used to control weeds (Berton, 1993). All this is in spite of what Marty Strange (1988) has called a 'stewardship penalty.' Farmers in the US who reduce their acreage base in grain crops risk losing valuable subsidies from the government under present agricultural policy. Still, families perceive the need for programs like Earth Husbandry (Chapter 12) and will, if given a chance, strengthen sustainable forms of production.

On other fronts family producers need to struggle to protect their place in the market. Threats from imports and other forms of production must be met by using the strengths of the family structure in cooperation with colleagues. Although some of the advantages of family agriculture may seem unclear to those who seek cheap food regardless of resource depletion, time is on our side. Information is accumulating on the real costs of water, the consequences of using large doses of chemicals and the social disturbances that result from labor saving machinery. The toll on farmers' health will be determined by the US Government, which is launching a ten year research project to determine why 'Farmers and people who work in their fields tend to have certain kinds of cancer more often than everyone else.' (The *Columbus Dispatch*, 14 February, 1993). Past research has shown that 'Farmers have non-Hodgkin's lymphoma, brain cancer and leukemia more often than the general population.... They also tend to be more prone ... to multiple myeloma and cancers of the brain, prostate, stomach, skin and lip.' We're paying a big price to participate in modern agriculture. It might be wise to stop and ponder the direction of things to come ...

We've looked over a broad territory and a range of topics. Hopefully the reader has developed some feeling for family agriculturalists in other parts of the world. Perhaps we now have a little more assurance that the joys and privileges of family agriculture will continue in the future. And maybe we even have an idea or two as to how we can lead our colleagues to understand more fully the significance of our profession.

Spread the word.

References

Almås, R (1985) New Forms of Cooperation in Norwegian Agriculture, in T Bergmann and T B Ogura (eds) *Cooperation in World Agriculture: Experiences, Problems and Perspectives* Tokyo: Food and Agriculture Policy Research Center, pp 43–57

Almaas, R (1991) Farm Policies and Farmer Strategies: The Case of Norway, in W H Friedland, L Busch, F H Buttle and A P Rudy (eds) *Towards a New Political Economy of Agriculture* Boulder, CO: Westview Press, pp 275–88

Altieri, M E (1990) Agroecology and Rural Development in Latin America, in M E Altieri and S B Hecht (eds) *Agroecology and Small Farm Development* Boca Raton: CRC Press, pp 113–120

Andersson, C (1985) *Peasant or Proletarian? Wage Labor and Peasant Economy during Industrialization, the Algerian Experience* Göteborg, Sweden: Almqvist & Wiksell International

Barker, R and D Chapman (1990) The Economics of Sustainable Agricultural Systems in Developing Countries, in C A Edwards, R Lal, P Madden, R H Miller and G House (eds) *Sustainable Agricultural Systems* Ankeny, IA: Soil and Water Conservation Society, pp 478–94

Barnes, D G (1930) *A History of the English Corn Laws* New York: Augustus M Kelley

Barrett, G W, N Rodenhouse and P J Bohlen (1990) Role of Sustainable Agriculture in Rural Landscapes, in C A Edwards, R Lal, P Madden, R H Miller and G House (eds) *Sustainable Agricultural Systems* Ankeny, IA: Soil and Water Conservation Society, pp 624–36

Barry, T (1987) *Roots of Rebellion, Land and Hunger in Central America* Boston MA: South End Press

Basseches, M (1984) *Dialecticial Thinking and Adult Development* New Jersey: Ablex Publishing

Bawden, R J (1991) Systems Thinking and Practice in Agriculture *Journal of Dairy Science* 74 (7): 2362–73

Beckford, G (1972) *Persistent Poverty, Underdevelopment in Plantation Economies of the Third World* New York: Oxford University Press

Benbrook, C M (1990) Society's Stake in Sustainable Agriculture, in C A Edwards, R Lal, P Madden, R H Miller and G House (eds)

Sustainable Agricultural Systems Ankeny, IA: Soil and Water Conservation Society, pp 68–76

Berton, V (1993) Earth Blankets, Winter Cover Crops Improve Soil, Stem Erosion *American Farmland* Washington DC: American Farmland Trust, pp 4–5

Blarel, B, P Hazell, F Place and J Quiggin (1992) The Economics of Farm Fragmentation: Evidence from Ghana and Rwanda *The World Bank Economic Review* 6 (2): 233–54

Boschwitz, R (1992) Foreword, in: W P Browne, J R Skees, L E Swanson, P B Thompson and L J Unnevehr *Sacred Cows and Hot Potatoes: Agrarian Myths in Agricultural Policy* Boulder, CO: Westview Press, pp xi–xii

Brady, N C (1990) Making Agriculture a Sustainable Industry, in C A Edwards, Rattan Lal et al *Sustainable Agricultural Systems* Ankeny, IA: Soil and Water Conservation Society, pp 20–32

Braus, P (1992) What Workers Want *American Demographics* Ithaca NY: American Demographics, 30–7

Brewster, D (1980) Changes in the Family Farm Concept, in *Farm Structure, a Historical Perspective on Changes in the Number and Size of Farms* Washington DC: Committee on Agriculture, Nutrition and Forestry, US Senate, pp 18–23

Brown, H C P and V G Thomas (1990) Ecological Considerations for the Future of Food Security in Africa, in C A Edwards, R Lal, P Madden, R H Miller and G House (eds) *Sustainable Agricultural Systems* Ankeny, IA: Soil and Water Conservation Society, pp 353–77

Browne, W P (1988) The Fragmented and Meandering Politics of Agriculture, in M A Tutwiler (ed) *U.S. Agriculture in a Global Setting: An Agenda for the Future* Washington DC: National Center for Food and Agricultural Policy, Annual Policy Review, 1987–1988 Resources for the Future, pp 136–53

Browne, W P, J R Skees, L E Swanson, P B Thompson and L J Unnevehr (1992) *Sacred Cows and Hot Potatoes, Agrarian Myths in Agricultural Policy.* Boulder, CO: Westview Press

Brum, A J (1988) *Modernização da Agriculture: Trigo e Soja* Petropolis, Rio de Janeiro, Brazil: Vozes

Busch, L and W B Lacy (1983) *Science, Agriculture, and the Politics of Research* Boulder, CO: Westview Press

Busch, L, W B Lacy, J Burkhardt and L M Lacy (1991) *Plants Power and Profit, Social, Economic, and Ethical Consequences of the New Biotechnologies* Cambridge, MA: Basil Blackwell

Buttel, F H, G W Gillespie Jr and A Power (1990) Sociological Aspects of Agricultural Sustainability in the United States: A New York Case Study, in C A Edwards, R Lal et al (eds) *Sustainable Agricultural Systems* Ankeny, IA: Soil and Water Conservation Society, pp 515–32

Chambers, R, Pacey, A and L A Thrupp (eds) (1989) *Farmer First, Farmer Innovation and Agricultural Research* London: Intermediate Technology Publications

Conway, G R and E B Barbier (1990) *After the Green Revolution: Sustainable Agriculture for Development* London: Earthscan Publications Ltd

Cornelius, J (1993) Stress and the Family Farm. Paper presented at the Centre for Agricultural Strategy/Small Farmers Association Symposium, University of Reading

Dahlberg, K A (1986) Introduction: Changing Contexts and Goals, in K A Dahlberg (Ed) *New Directions for Agriculture and Agricultural Research, Neglected Dimensions and Emerging Alternatives* Totowa, NJ: Rowman and Allanheld, pp 1–27

Davis, D and J Gaventa (1991) Altered States: Grassroots Movements and the Formation of Rural Policy. Paper presented at the national meeting of the *Rural Sociological Society*, Columbus, Ohio

Davis, J C (1986) *Rise from Want* Philadelphia PA: University of Pennsylvania Press

Edwards, C A (1990) The Importance of Integration in Sustainable Agricultural Systems, in C A Edwards, R Lal et al (eds) *Sustainable Agricultural Systems* Ankeny, IA: Soil and Water Conservation Society, pp 249–64

Edwards, C A, M K Wali, D J Horn and F Miller (1993) *Agriculture and the Environment* London: Elsevier Science Publishers

Egido, L G (1985) Agricultural Cooperation in Spain: The Agricultural Production in Common, in T Bergmann and T B Ogura (eds) *Cooperation in World Agriculture: Experience, Problems and Perspectives* Tokyo: Food and Agricultural Policy Research Center, pp 31–42

Flora, C B (1986) Values and the Agricultural Crisis: Differential Problems, Solutions and Value Constraints *Agriculture and Human Values* 3 (4): 16–23

Flora, C B (1990) The Social and Cultural Dynamics of Traditional Agricultural Communities, in M E Altieri and S B Hecht (eds) *Agroecology and Small Farm Development* Boca Raton: CRC Press, pp 27–34

Flora, C B and J L Flora (1988) Public Policy, Farm Size, and Community Well-Being in Farming-Dependent Counties of the Plains, in L E Swanson (Ed) *Agriculture and Community Change in the U.S.* Boulder, CO: Westview Press, pp 76–129

Friedland, W H (1991) Shaping the New Political Economy of Advanced Capitalist Agriculture, in W H Friedland, L Busch, F H Buttel, and A P Rudy (eds) *Towards a New Political Economy of Agriculture* Boulder, CO: Westview Press, pp 10034

Friedmann, H (1990) Family Wheat Farms and Third World Diets: A Paradoxical Relationship Between Unwaged and Waged Labor, in J

L Collins and M Gimenez (eds) *Work Without Wages, Comparative Studies of Domestic Labor and Self-Employment* Albany, NY: State University of New York Press, pp 193–213

Friedmann, H (1991) Changes in the International Division of Labor: Agri-food Complexes and Export Agriculture, in W H Friedland, L Busch, F H Buttel and A P Rudy (eds) *Towards a New Political Economy of Agriculture* Boulder, CO: Westview Press, pp 65–93

Fujimoto, I (1992) Lessons from Abroad in Rural Community Revitalization *Community Development Journal* 27 (1): pp. 10–20

Gliessman, R (1990) The Ecology and Management of Traditional Farming Systems, in M E Altieri and S B Hecht (eds) *Agroecology and Small Farm Development* Boca Raton: CRC Press, pp 13–8

Goodman, D (1991) Some Recent Tendencies in the Industrial Reorganization of the Agri-food System, in W H Friedland, L Busch et al (eds) *Towards a New Political Economy of Agriculture* Boulder, CO: Westview Press, pp 37–64

Gore, A (1992) *Earth in the Balance: Ecology and the Human Spirit* London: Earthscan Publications Ltd

Gyenes, A (1985) The Development of Farmers' Cooperatives in Hungary: Present Problems and Prospects, in T Bergmann and T B Ogura (eds) *Cooperation in World Agriculture: Experiences, Problems and Perspectives* Tokyo: Food and Agricultural Policy Research Center, pp 229–41

Hardin, G (1985) Lifeboat Ethics: The Case Against Helping the Poor, in G M Berardi (ed) *World Food, Population and Development* Totowa, NJ: Rowman and Allanheld, pp 108–15

Heffernan, W D (1986) Review and Evaluation of Social Externalities, in K A Dahlberg (ed) *New Directions for Agriculture and Agricultural Research, Neglected Dimensions and Emerging Alternatives* Totowa, NJ: Rowman and Allanheld, pp 199–220

Heiser, C B, Jr (1981) *Seeds to Civilization, the Story of Food* (2nd ed) San Francisco: W H Freeman & Co

Hildenbrand, B (1989) Tradition and Modernity in the Family Farm: A Case Study, in Boh, K, G Sgritta and M B Sussman (eds) *Cross-Cultural Perspectives on Families, Work and Change* Binghamton, NY: The Haworth Press, pp 159–72

Homer-Dixon, T F, J H Boutwell and G W Rathjens (1993) Environmental Change and Violent Conflict *Scientific American* 168 (2): 38–45

Institute for Community Economics (1982) *The Community Land Trust Handbook* Emmaus, PA: Rodale Press

International Food Policy Research Institute (1990) *Resource Transfers in the National and Provincial Economies, Agriculture and Economic Growth in Chile* Washington DC: IFPRI

Ka, C (1991) Agrarian Development, Family Farms and Sugar Capital in Colonial Taiwan, 1895–1945 *The Journal of Peasant Studies* 18 (2): 206–40

Kainz, H P (1988) *Paradox, Dialectic, and System: A Contemporary Reconstruction of the Hegelian Problematic* University Park PA: The Pennsylvania State University Press

Kelliher, D R (1985) *State Peasant Relations under China's Contemporary Reforms* (unpub PhD thesis) New Haven, CN: Yale University

Kenney, M, L M Lobao, J Curry, and W R Goe (1989) Midwestern Agriculture in U.S. Fordism: from the New Deal to Economic Restructuring *Sociologia Ruralis* 29 (2): 130–148

Kidron, M and R Segal (1991) *The New State of the World Atlas* (4th ed) New York: Simon and Schuster

Kirkby, R (1990) The Ecology of Traditional Agroecosystems in Africa, in M E Altieri and S B Hecht (eds) *Agroecology and Small Farm Development* Boca Raton: CRC Press, pp 173–82

Klicar, F M (1991) *The Yugoslav Republics* Downers Grove, IL: Megamark Film

Knutson, R D, J B Penn and W T Bochm (1983) *Agricultural and Food Policy* Englewood Cliffs, NJ: Prentice Hall

Kraenzel, C F (1980) *The Society Cost of Space in the Yonland* Bozeman MA: Big Sky Books, Montana State University

Lal, R, D J Eckert, N R Fausey and W M Edwards (1990) Conservation Tillage in Sustainable Agriculture, in C A Edwards, R Lal et al (eds) *Sustainable Agricultural Systems* Ankeny, IA: Soil and Water Conservation Society, pp 203–25

Lal, R and B A Stewart (1992) Need for Soil Restoration, in Lal, R and B A Stewart (eds) *Soil Restoration* New York: Springer-Verlag, pp 1–11

Levins, R and R Lewontin (1985) *The Dialectical Biologist* Cambridge MA: Harvard University Press

Lewis, G, A David, S Andréasson and P Allebeck (1992) Schizophrenia and City Life *The Lancet* 340: 137–40

Lobao, L M (1990) *Locality and Inequality: Farm and Industry Structure and Socioeconomic Conditions* Albany, NY: State University of New York Press

Luna, J M and G J House (1990) Pest Management in Sustainable Agricultural Systems, in C A Edwards, R Lal et al (eds) *Sustainable Agricultural Systems* Ankeny, IA: Soil and Water Conservation Society, pp 157–73

Luo, S and C Han (1990) Ecological Agriculture in China, in C A Edwards, R Lal et al (eds) *Sustainable Agricultural Systems* Ankeny, IA: Soil and Water Conservation Society, pp 299–322

Lyng, S (1990) *Holistic Health and Biomedical Medicine* Albany NY: State University of New York Press

MacCannell, D (1988) Industrial Agriculture and Rural Community

Degradation, in L E Swanson (ed) *Agriculture and Community Change in the U.S.* Boulder, CO: Westview Press, pp 15–75

Maybury-Lewis, D (1979) *Dialectical Societies; The Gê and Bororo of Central Brazil* Cambridge MA: Harvard University Press

McClellan, S (1991) Theorizing New Deal Farm Policy: Broad Constraints of Capital Accumulation and the Creation of a Hegemonic Relation, in W H Friedland, L Busch et al (eds) *Towards a New Political Economy of Agriculture* Boulder, CO: Westview Press, pp 215–31

Medina, J V (1977) *O Controle da Febre Aftosa nos Rebanhos de Paraguay* MS thesis, Federal University of Viçosa, Minas Gerais, Brazil

Mellor, J W (1988) Towards an Ethical Redistribution of Food and Agricultural Science, in LeMary, B W J (ed) *Science, Ethics, and Food* Washington DC: Smithsonian Institution Press, pp 78–92

Merrick, L (1990) Crop Diversity and its Conservation in Traditional Agroecosystems, in M E Altieri and S B Hecht (eds) *Agroecology and Small Farm Development* Boca Raton: CRC Press, pp 3–12

Moreira, M B (1991) Portuguese Agriculture and the State: An Outline of the Past 25 Years, in W H Friedland, L Busch et al (eds) *Towards a New Political Economy of Agriculture* Boulder, CO: Westview Press, pp 289–312

Mottura, G and E Mingione (1991) Agriculture and Agribusiness: Transformations and Trends in Italy, in W H Friedland, L Busch et al (eds) *Towards a New Political Economy of Agriculture* Boulder, CO; Westview Press, pp 94–112

Morgan, D (1985) Merchants of Grain (excerpts), in G M Berardi (ed) *World Food, Population and Development* Totowa, NJ: Rowman and Allanheld, pp 103–7

Murphy, R F (1971) *The Dialectics of Social Life: Alarms and Excursions in Anthropological Theory* New York: Basic Books

Ogura, T B and S Nakayasu (1985) Cooperative Organizations in Agricultural Production in Japan, in T Bergmann and T B Ogura (eds) *Cooperation in World Agriculture: Experiences, Problems and Perspectives* Tokyo: Food and Agricultural Policy Research Center, pp 72–80

Oshima, H T (1987) *Economic Growth in Monsoon Asia: A Comparative Study* Tokyo: University of Tokyo Press

Parr, J F, R I Papendic, I G Youngberg and R E Meyer (1990) Sustainable Agriculture in the United States, in C A Edwards, R Lal et al (eds) *Sustainable Agricultural Systems* Ankeny, IA: Soil and Water Conservation Society, pp 50–67

Penrose, R (1989) *The Emperor's New Mind: Concerning Computers, Minds, and the Laws of Physics* New York: Oxford University Press

Plucknett, D L (1990) International Goals and the Role of the International Agricultural Research Centers, in C A Edwards, R Lal et al (eds) *Sustainable Agricultural Systems* Ankeny, IA: Soil and Water Conservation Society, pp 33–49

Pretty, J N and R Howes (1993) Sustainable Agriculture in Britain: Recent Achievements and New Policy Challenges *International Institute for Environment and Development Research Series* London: IIED vol 2

Radwan, S and E Lee (1986) *Agrarian Change in Egypt: An Anatomy of Rural Poverty* London: Croom Helm

Rambaud, P (1985) The Invention of the Family Cooperation of Work in Rural France, in T Bergmann and T B Ogura (eds) *Cooperation in World Agriculture: Experiences, Problems and Perspectives* Tokyo: Food and Agricultural Policy Research Center, pp 11–29

Rogers, E M (1983) *Diffusion of Innovations* (third ed) Glencoe, UK: The Free Press

Ruttan, V W (1988) Scale, Size, Technology and Structure: A Personal Perspective, in L J Robison (ed) *Determinants of Farm Size and Structure* Michigan Agricultural Experiment Station Journal Article no 12899, pp 49–60

Ruttan, V W (1991) Constraints on Sustainable Growth in Agricultural Production: Into the 21st Century *Outlook on Agriculture* 20 (4): pp 225–34

Schneider, H (1984) *Meeting Food Needs in a Context of Change* Paris: Development Centre of the Organization for Economic Co-operation and Development

Sinclair, P R (1980) Agricultural Policy and the Decline of Commercial Family Farming: A Comparative Analysis of the U.S., Sweden, and the Netherlands, in F H Buttel and H Newby (eds) *The Rural Sociology of Advanced Societies* Montclair NJ: Allanheld, Osmun, pp 327–249

Stanton, B F (1989) Changes in Farm Size and Structure in American Agriculture in the Twentieth Century, in A Hallam (ed) *Determinants of Farm Size and Structure* Department of Economics, Iowa State University, pp 11–46

Stonich, S C (1991) The Political Economy of Environmental Destruction: Food Security in Southern Honduras, in M D Whiteford and A E Ferguson (eds) *Harvest of Want: Hunger and Food Security in Central America and Mexico* Boulder, CO: Westview Press, pp 45–74

Strange, M (1988) *Family Farming, a New Economic Vision* Lincoln NE: Institute for Food Development Policy, University of Nebraska

Swack, M (1987) Community Finance Institutions, in S T Bruyn and J Meehan (eds) *Beyond the Market and the State: New Directions for Community Development* Philadelphia: Temple University Press, pp 79–96

Swann, D (1988) *The Economics of the Common Market* London: Clays Ltd

Szelényi, I (1988) *Socialist Entrepreneurs, Embourgeoisement in Rural Hungary* Madison WI: University of Wisconsin Press

Thompson, P B (1986) The Social Goals of Agriculture *Agriculture and Human Values* 3 (4): 32–42

Thurston, A (1987) *Smallholder Agriculture in Colonial Kenya* Cambridge African Monographs 8, Cambridge, England: African Studies Centre

Tiffen, M and M Mortimore (1990) *Theory and Practice in Plantation Agriculture* Boulder, CO: Westview Press

Turner, C (1987) Worker Cooperatives and Community Development, in S T Bruyn and J Meehan (eds) *Beyond the Market and the State: New Directions for Community Development* Philadelphia PA: Temple University Press, pp 64–78

— (1988) The Philosophical Rationale for U.S. Agricultural Policy, in M A Tutwiler (ed) *U.S. Agriculture in a Global Setting: An Agenda for the Future* Washington DC: National Center for Food and Agricultural Policy, Annual Policy Review, 1987–1988. Resources for the Future, pp 34–45

Tutwiler, M A and B J Elliott (1988) An Interdependent and Fragile Global Economy, in M A Tutwiler (ed) *U.S. Agriculture in a Global Setting: An Agenda for the Future* Washington DC: National Center for Food and Agricultural Policy, Annual Policy Review, 1987–1988. Resources for the Future, pp 49–71

Tweeten, L (1989) *Farm Policy Analysis* Boulder, CO: Westview Press

United States Department of Agriculture (USDA) (1987) *The Farm Sector: How is it Weathering the 1980's?* AIB -506 Washington DC: Agriculture and Rural Economics Division

United States Department of Agriculture (USDA) (1990) *Alternative Opportunities in Agriculture: Expanding Output Through Diversification* Agricultural Economic Report no 633 Washington DC: Economic Research Service

US Congress (1991) *Rural America at the Crossroads: Networking for the Future* Washington DC: Office of Technology Assessment

Vail, D (1986) Crisis in Swedish Farmland Preservation Strategy *Agriculture and Human Values* 3 (4): 24–31

Vail, D (1991) Economic and Ecological Crises: Transforming Swedish Agricultural Policy, in W H Friedland, J Busch et al (eds) *Towards a New Political Economy of Agriculture* Boulder, CO: Westview Press, pp 256–74

Ward, N (1993) Environmental Concern and the Decline of Dynastic Family Farming. Paper presented at the Centre for Agricultural Strategy/Small Farmers Association Symposium, University of Reading, England

Watkins, W P (1986) *Cooperative Principles: Today and Tomorrow* Manchester: Holywake

Whatmore, S (1991) *Farming Women, Gender, Work and Family Enterprise* London: Macmillan

White, K and C Matthei (1987) Community Land Trusts, in S T Bruyn and J Meehan (eds) *Beyond the Market and the State: New Directions for Community Development* Philadelphia PA: Temple University Press, pp 41–64

Whiteford, M B (1991) From *Gallo Pinto* to 'Jack's Snacks': Observations on Dietary Change in a Rural Costa Rican Village, in M B Whiteford and A E Ferguson (eds) *Harvest of Want: Hunger and Food Security in Central America and Mexico* Boulder, CO: Westview Press, pp 127–40

Whiteford, M B and A E Ferguson (1991) Social Dimensions of Food Security and Hunger: An Overview, in M B Whiteford and A E Ferguson (eds) *Harvest of Want: Hunger and Food Security in Central America and Mexico* Boulder, CO: Westview Press, pp 1–21

World Resources Institute (1992) *World Resources, 1992–93* Oxford: Oxford University Press

Zdenek, R (1987) Community Development Corporations, in S T Bruyn and J Meehan (eds) *Beyond the Market and the State: New Directions for Community Development* Philadelphia PA: Temple University Press, pp 112–27

Index